FREE Test Taking Tips DVD Offer

To help us better serve you, we have developed a Test Taking Tips DVD that we would like to give you for FREE. **This DVD covers world-class test taking tips that you can use to be even more successful when you are taking your test.**

All that we ask is that you email us your feedback about your study guide. Please let us know what you thought about it – whether that is good, bad or indifferent.

To get your **FREE Test Taking Tips DVD**, email freedvd@studyguideteam.com with "FREE DVD" in the subject line and the following information in the body of the email:

  a. The title of your study guide.

  b. Your product rating on a scale of 1-5, with 5 being the highest rating.

  c. Your feedback about the study guide. What did you think of it?

  d. Your full name and shipping address to send your free DVD.

If you have any questions or concerns, please don't hesitate to contact us at freedvd@studyguideteam.com.

Thanks again!

OAR Study Guide 2019 & 2020

OAR Test Prep and Practice Test Questions for the Officer Aptitude Rating Exam
[Includes Detailed Answer Explanations]

Test Prep Books

Table of Contents

Quick Overview

As you draw closer to taking your exam, effective preparation becomes more and more important. Thankfully, you have this study guide to help you get ready. Use this guide to help keep your studying on track and refer to it often.

This study guide contains several key sections that will help you be successful on your exam. The guide contains tips for what you should do the night before and the day of the test. Also included are test-taking tips. Knowing the right information is not always enough. Many well-prepared test takers struggle with exams. These tips will help equip you to accurately read, assess, and answer test questions.

A large part of the guide is devoted to showing you what content to expect on the exam and to helping you better understand that content. In this guide are practice test questions so that you can see how well you have grasped the content. Then, answer explanations are provided so that you can understand why you missed certain questions.

Don't try to cram the night before you take your exam. This is not a wise strategy for a few reasons. First, your retention of the information will be low. Your time would be better used by reviewing information you already know rather than trying to learn a lot of new information. Second, you will likely become stressed as you try to gain a large amount of knowledge in a short amount of time. Third, you will be depriving yourself of sleep. So be sure to go to bed at a reasonable time the night before. Being well-rested helps you focus and remain calm.

Be sure to eat a substantial breakfast the morning of the exam. If you are taking the exam in the afternoon, be sure to have a good lunch as well. Being hungry is distracting and can make it difficult to focus. You have hopefully spent lots of time preparing for the exam. Don't let an empty stomach get in the way of success!

When travelling to the testing center, leave earlier than needed. That way, you have a buffer in case you experience any delays. This will help you remain calm and will keep you from missing your appointment time at the testing center.

Be sure to pace yourself during the exam. Don't try to rush through the exam. There is no need to risk performing poorly on the exam just so you can leave the testing center early. Allow yourself to use all of the allotted time if needed.

Remain positive while taking the exam even if you feel like you are performing poorly. Thinking about the content you should have mastered will not help you perform better on the exam.

Once the exam is complete, take some time to relax. Even if you feel that you need to take the exam again, you will be well served by some down time before you begin studying again. It's often easier to convince yourself to study if you know that it will come with a reward!

Test-Taking Strategies

1. Predicting the Answer

When you feel confident in your preparation for a multiple-choice test, try predicting the answer before reading the answer choices. This is especially useful on questions that test objective factual knowledge. By predicting the answer before reading the available choices, you eliminate the possibility that you will be distracted or led astray by an incorrect answer choice. You will feel more confident in your selection if you read the question, predict the answer, and then find your prediction among the answer choices. After using this strategy, be sure to still read all of the answer choices carefully and completely. If you feel unprepared, you should not attempt to predict the answers. This would be a waste of time and an opportunity for your mind to wander in the wrong direction.

2. Reading the Whole Question

Too often, test takers scan a multiple-choice question, recognize a few familiar words, and immediately jump to the answer choices. Test authors are aware of this common impatience, and they will sometimes prey upon it. For instance, a test author might subtly turn the question into a negative, or he or she might redirect the focus of the question right at the end. The only way to avoid falling into these traps is to read the entirety of the question carefully before reading the answer choices.

3. Looking for Wrong Answers

Long and complicated multiple-choice questions can be intimidating. One way to simplify a difficult multiple-choice question is to eliminate all of the answer choices that are clearly wrong. In most sets of answers, there will be at least one selection that can be dismissed right away. If the test is administered on paper, the test taker could draw a line through it to indicate that it may be ignored; otherwise, the test taker will have to perform this operation mentally or on scratch paper. In either case, once the obviously incorrect answers have been eliminated, the remaining choices may be considered. Sometimes identifying the clearly wrong answers will give the test taker some information about the correct answer. For instance, if one of the remaining answer choices is a direct opposite of one of the eliminated answer choices, it may well be the correct answer. The opposite of obviously wrong is obviously right! Of course, this is not always the case. Some answers are obviously incorrect simply because they are irrelevant to the question being asked. Still, identifying and eliminating some incorrect answer choices is a good way to simplify a multiple-choice question.

4. Don't Overanalyze

Anxious test takers often overanalyze questions. When you are nervous, your brain will often run wild, causing you to make associations and discover clues that don't actually exist. If you feel that this may be a problem for you, do whatever you can to slow down during the test. Try taking a deep breath or counting to ten. As you read and consider the question, restrict yourself to the particular words used by the author. Avoid thought tangents about what the author *really* meant, or what he or she was *trying* to say. The only things that matter on a multiple-choice test are the words that are actually in the question. You must avoid reading too much into a multiple-choice question, or supposing that the writer meant something other than what he or she wrote.

5. No Need for Panic

It is wise to learn as many strategies as possible before taking a multiple-choice test, but it is likely that you will come across a few questions for which you simply don't know the answer. In this situation, avoid panicking. Because most multiple-choice tests include dozens of questions, the relative value of a single wrong answer is small. As much as possible, you should compartmentalize each question on a multiple-choice test. In other words, you should not allow your feelings about one question to affect your success on the others. When you find a question that you either don't understand or don't know how to answer, just take a deep breath and do your best. Read the entire question slowly and carefully. Try rephrasing the question a couple of different ways. Then, read all of the answer choices carefully. After eliminating obviously wrong answers, make a selection and move on to the next question.

6. Confusing Answer Choices

When working on a difficult multiple-choice question, there may be a tendency to focus on the answer choices that are the easiest to understand. Many people, whether consciously or not, gravitate to the answer choices that require the least concentration, knowledge, and memory. This is a mistake. When you come across an answer choice that is confusing, you should give it extra attention. A question might be confusing because you do not know the subject matter to which it refers. If this is the case, don't eliminate the answer before you have affirmatively settled on another. When you come across an answer choice of this type, set it aside as you look at the remaining choices. If you can confidently assert that one of the other choices is correct, you can leave the confusing answer aside. Otherwise, you will need to take a moment to try to better understand the confusing answer choice. Rephrasing is one way to tease out the sense of a confusing answer choice.

7. Your First Instinct

Many people struggle with multiple-choice tests because they overthink the questions. If you have studied sufficiently for the test, you should be prepared to trust your first instinct once you have carefully and completely read the question and all of the answer choices. There is a great deal of research suggesting that the mind can come to the correct conclusion very quickly once it has obtained all of the relevant information. At times, it may seem to you as if your intuition is working faster even than your reasoning mind. This may in fact be true. The knowledge you obtain while studying may be retrieved from your subconscious before you have a chance to work out the associations that support it. Verify your instinct by working out the reasons that it should be trusted.

8. Key Words

Many test takers struggle with multiple-choice questions because they have poor reading comprehension skills. Quickly reading and understanding a multiple-choice question requires a mixture of skill and experience. To help with this, try jotting down a few key words and phrases on a piece of scrap paper. Doing this concentrates the process of reading and forces the mind to weigh the relative importance of the question's parts. In selecting words and phrases to write down, the test taker thinks about the question more deeply and carefully. This is especially true for multiple-choice questions that are preceded by a long prompt.

9. Subtle Negatives

One of the oldest tricks in the multiple-choice test writer's book is to subtly reverse the meaning of a question with a word like *not* or *except*. If you are not paying attention to each word in the question, you can easily be led astray by this trick. For instance, a common question format is, "Which of the following is...?" Obviously, if the question instead is, "Which of the following is not...?," then the answer will be quite different. Even worse, the test makers are aware of the potential for this mistake and will include one answer choice that would be correct if the question were not negated or reversed. A test taker who misses the reversal will find what he or she believes to be a correct answer and will be so confident that he or she will fail to reread the question and discover the original error. The only way to avoid this is to practice a wide variety of multiple-choice questions and to pay close attention to each and every word.

10. Reading Every Answer Choice

It may seem obvious, but you should always read every one of the answer choices! Too many test takers fall into the habit of scanning the question and assuming that they understand the question because they recognize a few key words. From there, they pick the first answer choice that answers the question they believe they have read. Test takers who read all of the answer choices might discover that one of the latter answer choices is actually *more* correct. Moreover, reading all of the answer choices can remind you of facts related to the question that can help you arrive at the correct answer. Sometimes, a misstatement or incorrect detail in one of the latter answer choices will trigger your memory of the subject and will enable you to find the right answer. Failing to read all of the answer choices is like not reading all of the items on a restaurant menu: you might miss out on the perfect choice.

11. Spot the Hedges

One of the keys to success on multiple-choice tests is paying close attention to every word. This is never truer than with words like almost, most, some, and sometimes. These words are called "hedges" because they indicate that a statement is not totally true or not true in every place and time. An absolute statement will contain no hedges, but in many subjects, the answers are not always straightforward or absolute. There are always exceptions to the rules in these subjects. For this reason, you should favor those multiple-choice questions that contain hedging language. The presence of qualifying words indicates that the author is taking special care with his or her words, which is certainly important when composing the right answer. After all, there are many ways to be wrong, but there is only one way to be right! For this reason, it is wise to avoid answers that are absolute when taking a multiple-choice test. An absolute answer is one that says things are either all one way or all another. They often include words like *every, always, best,* and *never*. If you are taking a multiple-choice test in a subject that doesn't lend itself to absolute answers, be on your guard if you see any of these words.

12. Long Answers

In many subject areas, the answers are not simple. As already mentioned, the right answer often requires hedges. Another common feature of the answers to a complex or subjective question are qualifying clauses, which are groups of words that subtly modify the meaning of the sentence. If the question or answer choice describes a rule to which there are exceptions or the subject matter is complicated, ambiguous, or confusing, the correct answer will require many words in order to be expressed clearly and accurately. In essence, you should not be deterred by answer choices that seem excessively long. Oftentimes, the author of the text will not be able to write the correct answer without offering some qualifications and modifications. Your job is to read the answer choices thoroughly and completely and to select the one that most accurately and precisely answers the question.

13. Restating to Understand

Sometimes, a question on a multiple-choice test is difficult not because of what it asks but because of how it is written. If this is the case, restate the question or answer choice in different words. This process serves a couple of important purposes. First, it forces you to concentrate on the core of the question. In order to rephrase the question accurately, you have to understand it well. Rephrasing the question will concentrate your mind on the key words and ideas. Second, it will present the information to your mind in a fresh way. This process may trigger your memory and render some useful scrap of information picked up while studying.

14. True Statements

Sometimes an answer choice will be true in itself, but it does not answer the question. This is one of the main reasons why it is essential to read the question carefully and completely before proceeding to the answer choices. Too often, test takers skip ahead to the answer choices and look for true statements. Having found one of these, they are content to select it without reference to the question above. Obviously, this provides an easy way for test makers to play tricks. The savvy test taker will always read the entire question before turning to the answer choices. Then, having settled on a correct answer choice, he or she will refer to the original question and ensure that the selected answer is relevant. The mistake of choosing a correct-but-irrelevant answer choice is especially common on questions related to specific pieces of objective knowledge. A prepared test taker will have a wealth of factual knowledge at his or her disposal, and should not be careless in its application.

15. No Patterns

One of the more dangerous ideas that circulates about multiple-choice tests is that the correct answers tend to fall into patterns. These erroneous ideas range from a belief that B and C are the most common right answers, to the idea that an unprepared test-taker should answer "A-B-A-C-A-D-A-B-A." It cannot be emphasized enough that pattern-seeking of this type is exactly the WRONG way to approach a multiple-choice test. To begin with, it is highly unlikely that the test maker will plot the correct answers according to some predetermined pattern. The questions are scrambled and delivered in a random order. Furthermore, even if the test maker was following a pattern in the assignation of correct answers, there is no reason why the test taker would know which pattern he or she was using. Any attempt to discern a pattern in the answer choices is a waste of time and a distraction from the real work of taking the test. A test taker would be much better served by extra preparation before the test than by reliance on a pattern in the answers.

FREE DVD OFFER

Don't forget that doing well on your exam includes both understanding the test content and understanding how to use what you know to do well on the test. We offer a completely FREE Test Taking Tips DVD that covers world class test taking tips that you can use to be even more successful when you are taking your test.

All that we ask is that you email us your feedback about your study guide. To get your **FREE Test Taking Tips DVD**, email freedvd@studyguideteam.com with "FREE DVD" in the subject line and the following information in the body of the email:

- The title of your study guide.
- Your product rating on a scale of 1-5, with 5 being the highest rating.
- Your feedback about the study guide. What did you think of it?
- Your full name and shipping address to send your free DVD.

Introduction to the OAR

Function of the Test

The Officer Aptitude Rating test is used by the United States Navy to determine how individuals will perform while in Naval Officer Candidate School (non-aviation). This test is part of the larger Aviation Selection Test Battery (ASTB), which is used to select individuals for flight and pilot officer training programs offered by the U.S. Coast Guard, the U.S. Marine Corps, and the U.S. Navy.

Test Administration

Candidates can take the OAR test at over 250 registered locations all over the world. These locations include naval officer recruiting stations, military institutes, and NROTC units at major universities. Officer recruiters schedule test dates for exam candidates once examinees prove that they meet the qualifications. The exam is offered in two different formats: a paper form and a computer adaptive (online) exam via the web-based APEX system. No two online tests are identical due to the fact that they are automatically generated from a library containing hundreds of potential questions for each of the three subtests.

Test takers who wish to retake the OAR test must wait thirty days in order to do so, and they can only take the OAR test a total of three times over the course of their lifetime. It is important to note that a test taker's most recent test score replaces any OAR score on record, even if he or she received a higher score on a previous test. Test scores are valid for life.

Test Format

The Math Skills Test (MST) utilizes both word problems and equations to test examinees on high school math concepts dealing with algebra and geometry. The Reading Comprehension Test (RCT) incorporates text passages to test examinees on their ability to extract information and make conclusions from what they have read. Finally, the Mechanical Comprehension Test (MCT) is comprised of questions involving high school physics concepts, as well simple machine mechanics.

The following table contains a breakdown of the number of questions and the corresponding time limits for the various subtests of the OAR exam for examinees that choose to take the paper version:

Sections of the OAR Test – Paper Version		
Subject Areas	Number of Questions (Multiple-Choice)	Time Limit
Math Skills Test	30	40 minutes
Reading Comprehension Test	20	30 minutes
Mechanical Comprehension Test	30	15 minutes
Total	80	85 minutes

For test takers who choose to take the online version of the exam, the test can ninety minutes to two hours to complete. There are between twenty and thirty multiple-choice questions within each subtest.

Scoring

Individuals taking the paper version of the OAR test are penalized for leaving questions unanswered, and thus, they are best served by making educated guesses when time is running out or the answer is unknown. In contrast, individuals who are taking the online version of the OAR will not benefit from guessing. It is more important to work accurately and quickly on this version of the exam.

The OAR test is scored on a scale of 20-80 in increments of one point, and a minimum score of 35 is needed to pass. However, most individuals who take the exam receive scores between 40-60. There has been no real difference in passing rates between test takers who take the paper version of the exam versus those who take the online version.

Test takers who attempt the online version of the exam receive their scores immediately upon completion of the exam, while those who take the paper version must send their completed tests into Navy Medicine Operational Training Center (NMOTC) to be scored.

Recent/Future Developments

At the end of 2013, a new version of the larger Aviation Selection Test Battery (ASTB) was released, which is known as the ASTB-E. Three of the seven subtests still make up the OAR test. The overall exam is now better able to gauge how test takers think in different dimensions (hand-eye coordination, physical dexterity, and dividing attention between tasks) through what is known as the Performance Based Measures Battery.

Math Skills

Arithmetic

How to Prepare

These problems involve basic arithmetic skills as well as the ability to break down a word problem to see where to apply these skills in order to get the correct answer. The basics of arithmetic and the approach to solving word problems are discussed here.

Note that math requires practice in order to become proficient. Make sure to not just read through the material here, but also try out the practice questions, as well as check the answers provided. Just reading through examples does not necessarily mean that a student can do the problems themselves. Note that sometimes there can be multiple approaches to getting a solution when doing the problems. What matters is getting the correct answer, so it is okay if the approach to a problem is different than the solution method provided.

Basic Operations of Arithmetic

There are four different basic operations used with numbers: addition, subtraction, multiplication, and division.

- Addition takes two numbers and combines them into a total called the sum. The sum is the total when combining two collections into one. If there are 5 things in one collection and 3 in another, then after combining them, there is a total of $5 + 3 = 8$. Note the order does not matter when adding numbers. For example, $3 + 5 = 8$.

- Subtraction is the opposite (or "inverse") operation to addition. Whereas addition combines two quantities together, subtraction takes one quantity away from another. For example, if there are 20 gallons of fuel and 5 are removed, that gives $20 - 5 = 15$ gallons remaining. Note that for subtraction, the order does matter because it makes a difference which quantity is being removed from which.

- Multiplication is repeated addition. 3×4 can be thought of as putting together 3 sets of items, each set containing 4 items. The total is 12 items. Another way to think of this is to think of each number as the length of one side of a rectangle. If a rectangle is covered in tiles with 3 columns of 4 tiles each, then there are 12 tiles in total. From this, one can see that the answer is the same if the rectangle has 4 rows of 3 tiles each: $4 \times 3 = 12$. By expanding this reasoning, the order the numbers are multiplied does not matter.

- Division is the opposite of multiplication. It means taking one quantity and dividing it into sets the size of the second quantity. If there are 16 sandwiches to be distributed to 4 people, then each person gets $16 \div 4 = 4$ sandwiches. As with subtraction, the order in which the numbers appear does matter for division.

Addition

Addition is the combination of two numbers so their quantities are added together cumulatively. The sign for an addition operation is the + symbol. For example, $9 + 6 = 15$. The 9 and 6 combine to achieve a cumulative value, called a sum.

Addition holds the commutative property, which means that numbers in an addition equation can be switched without altering the result. The formula for the commutative property is $a + b = b + a$. Let's look at a few examples to see how the commutative property works:

$$7 = 3 + 4 = 4 + 3 = 7$$

$$20 = 12 + 8 = 8 + 12 = 20$$

Addition also holds the associative property, which means that the grouping of numbers doesn't matter in an addition problem. In other words, the presence or absence of parentheses is irrelevant. The formula for the associative property is $(a + b) + c = a + (b + c)$. Here are some examples of the associative property at work:

$$30 = (6 + 14) + 10 = 6 + (14 + 10) = 30$$

$$35 = 8 + (2 + 25) = (8 + 2) + 25 = 35$$

Subtraction

Subtraction is taking away one number from another, so their quantities are reduced. The sign designating a subtraction operation is the − symbol, and the result is called the difference. For example, $9 - 6 = 3$. The number *6* detracts from the number *9* to reach the difference *3*.

Unlike addition, subtraction follows neither the commutative nor associative properties. The order and grouping in subtraction impact the result.

$$15 = 22 - 7 \neq 7 - 22 = -15$$

$$3 = (10 - 5) - 2 \neq 10 - (5 - 2) = 7$$

When working through subtraction problems involving larger numbers, it's necessary to regroup the numbers. Let's work through a practice problem using regrouping:

$$
\begin{array}{r}
3\ 2\ 5 \\
-\ 7\ 7 \\
\hline
\end{array}
$$

Here, it is clear that the ones and tens columns for 77 are greater than the ones and tens columns for 325. To subtract this number, borrow from the tens and hundreds columns. When borrowing from a column, subtracting 1 from the lender column will add 10 to the borrower column:

$$
\begin{array}{ccc}
3\text{-}1 & 10\text{+}2\text{-}1 & 10\text{+}5 \\
 & 7 & 7 \\
\hline
\end{array}
\quad = \quad
\begin{array}{ccc}
2 & 11 & 15 \\
 & 7 & 7 \\
\hline
2 & 4 & 8
\end{array}
$$

After ensuring that each digit in the top row is greater than the digit in the corresponding bottom row, subtraction can proceed as normal, and the answer is found to be 248.

Multiplication

Multiplication involves adding together multiple copies of a number. It is indicated by an × symbol or a number immediately outside of a parenthesis. For example:

$$5(8 - 2)$$

The two numbers being multiplied together are called factors, and their result is called a product. For example, $9 \times 6 = 54$. This can be shown alternatively by expansion of either the 9 or the 6:

$$9 \times 6 = 9 + 9 + 9 + 9 + 9 + 9 = 54$$

$$9 \times 6 = 6 + 6 + 6 + 6 + 6 + 6 + 6 + 6 + 6 = 54$$

Like addition, multiplication holds the commutative and associative properties:

$$115 = 23 \times 5 = 5 \times 23 = 115$$

$$84 = 3 \times (7 \times 4) = (3 \times 7) \times 4 = 84$$

Multiplication also follows the distributive property, which allows the multiplication to be distributed through parentheses. The formula for distribution is $a \times (b + c) = ab + ac$. This is clear after the examples:

$$45 = 5 \times 9 = 5(3 + 6) = (5 \times 3) + (5 \times 6) = 15 + 30 = 45$$

$$20 = 4 \times 5 = 4(10 - 5) = (4 \times 10) - (4 \times 5) = 40 - 20 = 20$$

Multiplication becomes slightly more complicated when multiplying numbers with decimals. The easiest way to answer these problems is to ignore the decimals and multiply as if they were whole numbers. After multiplying the factors, place a decimal in the product. The placement of the decimal is determined by taking the cumulative number of decimal places in the factors.

For example:

0.7	2.6	1.5
x 3	x 4.2	x 6.4
2.1	10.92	9.60

Let's tackle the first example. First, ignore the decimal and multiply the numbers as though they were whole numbers to arrive at a product: 21. Second, count the number of digits that follow a decimal (one). Finally, move the decimal place that many positions to the left, as the factors have only one decimal place. The second example works the same way, except that there are two total decimal places in the factors, so the product's decimal is moved two places over. In the third example, the decimal should be moved over two digits, but the digit zero is no longer needed, so it is erased and the final answer is 9.6.

Division

Division and multiplication are inverses of each other in the same way that addition and subtraction are opposites. The signs designating a division operation are the $\div$ and / symbols. In division, the second number divides into the first.

The number before the division sign is called the dividend or, if expressed as a fraction, the numerator. For example, in $a \div b$, a is the dividend, while in $\frac{a}{b}$, a is the numerator.

The number after the division sign is called the divisor or, if expressed as a fraction, the denominator. For example, in $a \div b$, b is the divisor, while in $\frac{a}{b}$, b is the denominator.

Like subtraction, division doesn't follow the commutative property, as it matters which number comes before the division sign, and division doesn't follow the associative or distributive properties for the same reason. For example:

$$\frac{3}{2} = 9 \div 6 \neq 6 \div 9 = \frac{2}{3}$$

$$2 = 10 \div 5 = (30 \div 3) \div 5 \neq 30 \div (3 \div 5) = 30 \div \frac{3}{5} = 50$$

$$25 = 20 + 5 = (40 \div 2) + (40 \div 8) \neq 40 \div (2 + 8) = 40 \div 10 = 4$$

If a divisor doesn't divide into a dividend an integer number of times, whatever is left over is termed the remainder. The remainder can be further divided out into decimal form by using long division; however, this doesn't always give a quotient with a finite number of decimal places, so the remainder can also be expressed as a fraction over the original divisor.

Division with decimals is similar to multiplication with decimals in that when dividing a decimal by a whole number, ignore the decimal and divide as if it were a whole number.

Upon finding the answer, or quotient, place the decimal at the decimal place equal to that in the dividend.

$$15.75 \div 3 = 5.25$$

When the divisor is a decimal number, multiply both the divisor and dividend by 10. Repeat this until the divisor is a whole number, then complete the division operation as described above.

$$17.5 \div 2.5 = 175 \div 25 = 7$$

Fractions

A *fraction* is a number used to express a ratio. It is written as a number x over a line with another number y underneath: $\frac{x}{y}$, and can be thought of as x out of y equal parts. The number on top (x) is called the *numerator*, and the number on the bottom is called the *denominator* (y). It is important to remember the only restriction is that the denominator is not allowed to be 0.

Imagine that an apple pie has been baked for a holiday party, and the full pie has eight slices. After the party, there are five slices left. How could the amount of the pie that remains be expressed as a fraction? The numerator is 5 since there are 5 pieces left, and the denominator is 8 since there were eight total slices in the whole pie. Thus, expressed as a fraction, the leftover pie totals $\frac{5}{8}$ of the original amount.

Another way of thinking about fractions is like this: $\frac{x}{y} = x \div y$.

Two fractions can sometimes equal the same number even when they look different. The value of a fraction will remain equal when multiplying both the numerator and the denominator by the same number. The value of the fraction does not change when dividing both the numerator and the denominator by the same number. For example, $\frac{4}{8} = \frac{2}{4} = \frac{1}{2}$. If two fractions look different, but are actually the same number, these are *equivalent fractions*.

A number that can divide evenly into a second number is called a *divisor* or *factor* of that second number; 3 is a divisor of 6, for example. If the numerator and denominator in a fraction have no common factors other than 1, the fraction is said to be *simplified*. $\frac{2}{4}$ is not simplified (since the numerator and denominator have a factor of 2 in common), but $\frac{1}{2}$ is simplified. Often, when solving a problem, the final answer generally requires us to simplify the fraction.

It is often useful when working with fractions to rewrite them so they have the same denominator. This process is called finding a *common denominator*. The common denominator of two fractions needs to be a number that is a multiple of both denominators. For example, given $\frac{1}{6}$ and $\frac{5}{8}$, a common denominator is $6 \times 8 = 48$. However, there are often smaller choices for the common denominator. The smallest number that is a multiple of two numbers is called the *least common multiple* of those numbers. For this example, use the numbers 6 and 8. The multiples of 6 are 6, 12, 18, 24... and the multiples of 8 are 8, 16, 24..., so the least common multiple is 24. The two fractions are rewritten as $\frac{4}{24}, \frac{15}{24}$.

If two fractions have a common denominator, then the numerators can be added or subtracted. For example, $\frac{4}{5} - \frac{3}{5} = \frac{4-3}{5} = \frac{1}{5}$. If the fractions are not given with the same denominator, a common denominator needs to be found before adding or subtracting them.

It is always possible to find a common denominator by multiplying the denominators by each other. However, when the denominators are large numbers, this method is unwieldy, especially if the answer must be provided in its simplest form. Thus, it's beneficial to find the least common denominator of the fractions—the least common denominator is incidentally also the least common multiple.

Once equivalent fractions have been found with common denominators, simply add or subtract the numerators to arrive at the answer:

$$1) \; \frac{1}{2} + \frac{3}{4} = \frac{2}{4} + \frac{3}{4} = \frac{5}{4}$$

$$2) \; \frac{3}{12} + \frac{11}{20} = \frac{15}{60} + \frac{33}{60} = \frac{48}{60} = \frac{4}{5}$$

$$3) \; \frac{7}{9} - \frac{4}{15} = \frac{35}{45} - \frac{12}{45} = \frac{23}{45}$$

$$4) \; \frac{5}{6} - \frac{7}{18} = \frac{15}{18} - \frac{7}{18} = \frac{8}{18} = \frac{4}{9}$$

One of the most fundamental concepts of fractions is their ability to be manipulated by multiplication or division. This is possible since $\frac{n}{n} = 1$ for any non-zero integer. As a result, multiplying or dividing by $\frac{n}{n}$ will not alter the original fraction since any number multiplied or divided by 1 doesn't change the value of that number. Fractions of the same value are known as equivalent fractions. For example, $\frac{2}{4}, \frac{4}{8}, \frac{50}{100},$ and $\frac{75}{150}$ are equivalent, as they all equal $\frac{1}{2}$.

To multiply two fractions, multiply the numerators to get the new numerator as well as multiply the denominators to get the new denominator. For example:

$$\frac{3}{5} \times \frac{2}{7} = \frac{3 \times 2}{5 \times 7} = \frac{6}{35}$$

Switching the numerator and denominator is called taking the *reciprocal* of a fraction. So the reciprocal of $\frac{4}{5}$ is $\frac{5}{4}$.

To divide one fraction by another, multiply the first fraction by the reciprocal of the second. So:

$$\frac{3}{4} \div \frac{2}{5} = \frac{3}{4} \times \frac{5}{2} = \frac{15}{8}$$

If the numerator is smaller than the denominator, the fraction is a *proper fraction*. Otherwise, the fraction is said to be *improper*.

A *mixed number* is a number that is an integer plus some proper fraction, and is written with the integer first and the proper fraction to the right of it. Any mixed number can be written as an improper fraction by multiplying the integer by the denominator, adding the product to the value of the numerator, and dividing the sum by the original denominator. For example:

$$3\frac{1}{2} = \frac{3 \times 2 + 1}{2} = \frac{7}{2}$$

Whole numbers can also be converted into fractions by placing the whole number as the numerator and making the denominator 1. For example, $3 = \frac{3}{1}$.

Percentages

Think of percentages as fractions with a denominator of 100. In fact, percentage means "per hundred." Problems often require converting numbers from percentages, fractions, and decimals. The following explains how to work through those conversions.

Converting Fractions to Percentages: Convert the fraction by using an equivalent fraction with a denominator of 100. For example:

$$\frac{3}{4} = \frac{3}{4} \times \frac{25}{25}$$

$$\frac{75}{100} = 75\%$$

Converting Percentages to Fractions: Percentages can be converted to fractions by turning the percentage into a fraction with a denominator of 100. Be wary of questions asking the converted fraction to be written in the simplest form. For example, $35\% = \frac{35}{100}$ which, although correctly written, has a numerator and denominator with a greatest common factor of 5 and can be simplified to $\frac{7}{20}$.

Converting Percentages to Decimals: As a percentage is based on "per hundred," decimals and percentages can be converted by multiplying or dividing by 100. Practically speaking, this always amounts to moving the decimal point two places to the right or left, depending on the conversion. To convert a percentage to a decimal, move the decimal point two places to the left and remove the % sign. To convert a decimal to a percentage, move the decimal point two places to the right and add a "%" sign. Here are some examples:

65% = 0.65
0.33 = 33%
0.215 = 21.5%
99.99% = 0.9999
500% = 5.00
7.55 = 755%

Questions dealing with percentages can be difficult when they are phrased as word problems. These word problems almost always come in three varieties. The first type will ask to find what percentage of some number will equal another number. The second asks to determine what number is some percentage of another given number. The third will ask what number another number is a given percentage of.

One of the most important parts of correctly answering percentage word problems is to identify the numerator and the denominator. This fraction can then be converted into a percentage, as described above.

The following word problem shows how to make this conversion:

> A department store carries several different types of footwear. The store is currently selling 8 athletic shoes, 7 dress shoes, and 5 sandals. What percentage of the store's footwear are sandals?

First, calculate what serves as the "whole," as this will be the denominator. How many total pieces of footwear does the store sell? The store sells 20 different types (8 athletic + 7 dress + 5 sandals).

Second, what footwear type is the question specifically asking about? Sandals. Thus, 5 is the numerator.

Third, the resultant fraction must be expressed as a percentage. The first two steps indicate that $\frac{5}{20}$ of the footwear pieces are sandals. This fraction must now be converted into a percentage:

$$\frac{5}{20} \times \frac{5}{5} = \frac{25}{100} = 25\%$$

Ratios and Proportions

Ratios are used to show the relationship between two quantities. The ratio of oranges to apples in the grocery store may be 3 to 2. That means that for every 3 oranges, there are 2 apples. This comparison can be expanded to represent the actual number of oranges and apples. Another example may be the number of boys to girls in a math class. If the ratio of boys to girls is given as 2 to 5, that means there are 2 boys to every 5 girls in the class. Ratios can also be compared if the units in each ratio are the same. The ratio of boys to girls in the math class can be compared to the ratio of boys to girls in a science class by stating which ratio is higher and which is lower.

Rates are used to compare two quantities with different units. *Unit rates* are the simplest form of rate. With unit rates, the denominator in the comparison of two units is one. For example, if someone can type at a rate of 1000 words in 5 minutes, then his or her unit rate for typing is $\frac{1000}{5} = 200$ words in one minute or 200 words per minute. Any rate can be converted into a unit rate by dividing to make the denominator one. 1000 words in 5 minutes has been converted into the unit rate of 200 words per minute.

Ratios and rates can be used together to convert rates into different units. For example, if someone is driving 50 kilometers per hour, that rate can be converted into miles per hour by using a ratio known as the *conversion factor*. Since the given value contains kilometers and the final answer needs to be in miles, the ratio relating miles to kilometers needs to be used. There are 0.62 miles in 1 kilometer. This, written as a ratio and in fraction form, is

$$\frac{0.62 \; miles}{1 \; km}$$

To convert 50km/hour into miles per hour, the following conversion needs to be set up:

$$\frac{50 \; km}{hour} \times \frac{0.62 \; miles}{1 \; km} = 31 \; miles \; per \; hour$$

The ratio between two similar geometric figures is called the *scale factor*. For example, a problem may depict two similar triangles, A and B. The scale factor from the smaller triangle A to the larger triangle B is given as 2 because the length of the corresponding side of the larger triangle, 16, is twice the corresponding side on the smaller triangle, 8. This scale factor can also be used to find the value of a missing side, x, in triangle A. Since the scale factor from the smaller triangle (A) to larger one (B) is 2, the larger corresponding side in triangle B (given as 25), can be divided by 2 to find the missing side in A ($x = 12.5$). The scale factor can also be represented in the equation $2A = B$ because two times the lengths of A gives the corresponding lengths of B. This is the idea behind similar triangles.

Much like a scale factor can be written using an equation like $2A = B$, a *relationship* is represented by the equation $Y = kX$. X and Y are proportional because as values of X increase, the values of Y also increase. A relationship that is inversely proportional can be represented by the equation $Y = \frac{k}{X}$, where the value of Y decreases as the value of x increases and vice versa.

Proportional reasoning can be used to solve problems involving ratios, percentages, and averages. Ratios can be used in setting up proportions and solving them to find unknowns. For example, if a student completes an average of 10 pages of math homework in 3 nights, how long would it take the student to complete 22 pages? Both ratios can be written as fractions. The second ratio would contain the unknown.

The following proportion represents this problem, where x is the unknown number of nights:

$$\frac{10 \ pages}{3 \ nights} = \frac{22 \ pages}{x \ nights}$$

Solving this proportion entails cross-multiplying and results in the following equation: $10x = 22 \times 3$. Simplifying and solving for x results in the exact solution: $x = 6.6 \ nights$. The result would be rounded up to 7 because the homework would actually be completed on the 7th night.

The following problem uses ratios involving percentages:

If 20% of the class is girls and 30 students are in the class, how many girls are in the class?

To set up this problem, it is helpful to use the common proportion:

$$\frac{\%}{100} = \frac{is}{of}$$

Within the proportion, % is the percentage of girls, 100 is the total percentage of the class, *is* is the number of girls, and *of* is the total number of students in the class. Most percentage problems can be written using this language. To solve this problem, the proportion should be set up as $\frac{20}{100} = \frac{x}{30}$, and then solved for x. Cross-multiplying results in the equation $20 \times 30 = 100x$, which results in the solution $x = 6$. There are 6 girls in the class.

Problems involving volume, length, and other units can also be solved using ratios. A problem may ask for the volume of a cone to be found that has a radius, $r = 7m$ and a height, $h = 16m$. Referring to the formulas provided on the test, the volume of a cone is given as:

$$V = \pi r^2 \frac{h}{3}$$

17

r is the radius, and h is the height. Plugging $r = 7$ and $h = 16$ into the formula, the following is obtained:

$$V = \pi(7^2)\frac{16}{3}$$

Therefore, volume of the cone is found to be approximately 821m³. Sometimes, answers in different units are sought. If this problem wanted the answer in liters, 821m³ would need to be converted.

Using the equivalence statement 1m³ = 1000L, the following ratio would be used to solve for liters:

$$821\mathrm{m}^3 \times \frac{1000L}{1m^3}$$

Cubic meters in the numerator and denominator cancel each other out, and the answer is converted to 821,000 liters, or 8.21×10^5 L.

Other conversions can also be made between different given and final units. If the temperature in a pool is 30°C, what is the temperature of the pool in degrees Fahrenheit? To convert these units, an equation is used relating Celsius to Fahrenheit. The following equation is used:

$$T_{°F} = 1.8T_{°C} + 32$$

Plugging in the given temperature and solving the equation for T yields the result:

$$T_{°F} = 1.8(30) + 32 = 86°F$$

Both units in the metric system and U.S. customary system are widely used.

Basic Geometry Relationships

The basic unit of geometry is a point. A point represents an exact location on a plane, or flat surface. The position of a point is indicated with a dot and usually named with a single uppercase letter, such as point *A* or point *T*. A point is a place, not a thing, and therefore has no dimensions or size. A set of points that lies on the same line is called collinear. A set of points that lies on the same plane is called coplanar.

B

C

A

The image above displays point *A*, point *B*, and point *C*.

A line is as series of points that extends in both directions without ending. It consists of an infinite number of points and is drawn with arrows on both ends to indicate it extends infinitely. Lines can be named by two points on the line or with a single, cursive, lower case letter. The two lines below could be named line *AB* or line *BA* or $\overleftrightarrow{AB}$ or $\overleftrightarrow{BA}$; and line *m*.

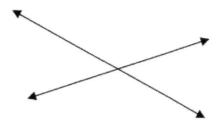

Two lines are considered parallel to each other if, while extending infinitely, they will never intersect (or meet). Parallel lines point in the same direction and are always the same distance apart. Two lines are considered perpendicular if they intersect to form right angles. Right angles are 90°. Typically, a small box is drawn at the intersection point to indicate the right angle.

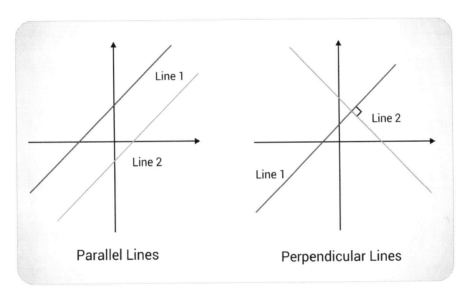

Line 1 is parallel to line 2 in the left image and is written as line 1 || line 2. Line 1 is perpendicular to line 2 in the right image and is written as line 1 ⊥ line 2.

A ray has a specific starting point and extends in one direction without ending. The endpoint of a ray is its starting point. Rays are named using the endpoint first, and any other point on the ray. The following ray can be named ray *AB* and written $\overrightarrow{AB}$.

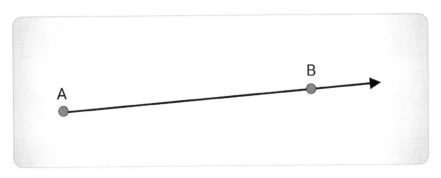

A line segment has specific starting and ending points. A line segment consists of two endpoints and all the points in between. Line segments are named by the two endpoints. The example below is named segment *KL* or segment *LK*, written $\overline{KL}$ or $\overline{LK}$.

Classification of Angles

An angle consists of two rays that have a common endpoint. This common endpoint is called the vertex of the angle. The two rays can be called sides of the angle. The angle below has a vertex at point *B* and the sides consist of ray *BA* and ray *BC*. An angle can be named in three ways:

- Using the vertex and a point from each side, with the vertex letter in the middle.
- Using only the vertex. This can only be used if it is the only angle with that vertex.
- Using a number that is written inside the angle.

The angle below can be written $\angle ABC$ (read angle *ABC*), $\angle CBA$, $\angle B$, or $\angle 1$.

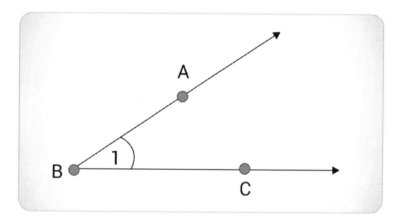

An angle divides a plane, or flat surface, into three parts: the angle itself, the interior (inside) of the angle, and the exterior (outside) of the angle. The figure below shows point *M* on the interior of the angle and point *N* on the exterior of the angle.

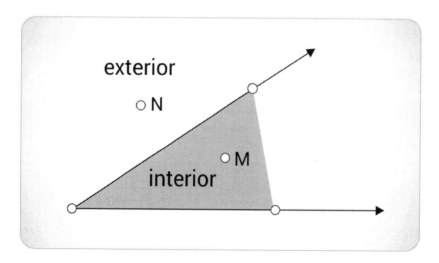

Angles can be measured in units called degrees, with the symbol °. The degree measure of an angle is between 0° and 180° and can be obtained by using a protractor.

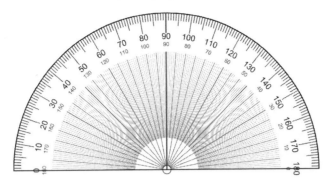

A straight angle (or simply a line) measures exactly 180°. A right angle's sides meet at the vertex to create a square corner. A right angle measures exactly 90° and is typically indicated by a box drawn in the interior of the angle. An acute angle has an interior that is narrower than a right angle. The measure of an acute angle is any value less than 90° and greater than 0°. For example, 89.9°, 47°, 12°, and 1°. An obtuse angle has an interior that is wider than a right angle. The measure of an obtuse angle is any value greater than 90° but less than 180°. For example, 90.1°, 110°, 150°, and 179.9°.

- Acute angles: Less than 90°
- Obtuse angles: Greater than 90°
- Right angles: 90°
- Straight angles: 180°

If two angles add together to give 90°, the angles are *complementary*.

If two angles add together to give 180°, the angles are *supplementary*.

When two lines intersect, the pairs of angles they form are always supplementary. The two angles marked here are supplementary:

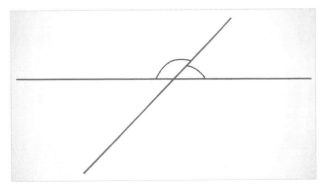

When two supplementary angles are next to one another or "adjacent" in this way, they always give rise to a straight line.

A polygon is a closed geometric figure in a plane (flat surface) consisting of at least 3 sides formed by line segments. These are often defined as two-dimensional shapes. Common two-dimensional shapes include circles, triangles, squares, rectangles, pentagons, and hexagons. Note that a circle is a two-dimensional shape without sides.

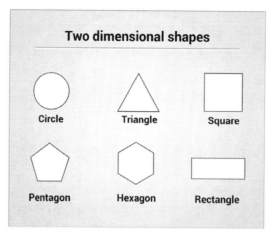

Polygons can be classified by the number of sides (also equal to the number of angles) they have. The following are the names of polygons with a given number of sides or angles:

# of sides	3	4	5	6	7	8	9	10
Name of polygon	Triangle	Quadrilateral	Pentagon	Hexagon	Septagon (or heptagon)	Octagon	Nonagon	Decagon

Triangles can be further classified by their sides and angles. A triangle with its largest angle measuring 90° is a right triangle. A triangle with the largest angle less than 90° is an acute triangle. A triangle with the largest angle greater than 90° is an obtuse triangle. Below is an example of a right triangle.

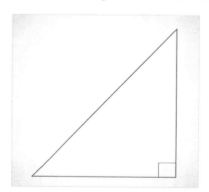

A triangle consisting of two equal sides and two equal angles is an isosceles triangle. A triangle with three equal sides and three equal angles is an equilateral triangle. A triangle with no equal sides or angles is a scalene triangle.

The three angles inside the triangle are called *interior angles* and add up to 180°.

For any triangle, the *Triangle Inequality Theorem* says that the following holds true: $A + B > C, A + C > B, B + C > A$. In addition, the sum of two angles must be less than 180°.

22

If two triangles have angles that agree with one another, that is, the angles of the first triangle are equal to the angles of the second triangle, then the triangles are called *similar*. Similar triangles look the same, but one can be a "magnification" of the other.

Two triangles with sides that are the same length must also be similar triangles. In this case, such triangles are called *congruent*. Congruent triangles have the same angles and lengths, even if they are rotated from one another.

Quadrilaterals can be further classified according to their sides and angles. A quadrilateral with exactly one pair of parallel sides is called a trapezoid. A quadrilateral that shows both pairs of opposite sides parallel is a parallelogram. Parallelograms include rhombuses, rectangles, and squares. A rhombus has four equal sides. A rectangle has four equal angles (90° each). A square has four 90° angles and four equal sides. Therefore, a square is both a rhombus and a rectangle.

A solid figure, or simply solid, is a figure that encloses a part of space. Some solids consist of flat surfaces only while others include curved surfaces. Solid figures are often defined as three-dimensional shapes. Common three-dimensional shapes include spheres, prisms, cubes, pyramids, cylinders, and cones.

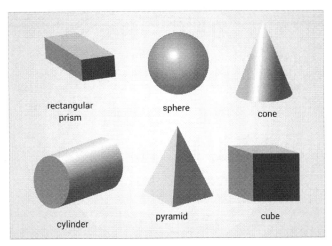

Perimeter is the measurement of a distance around something. It can be thought of as the length of the boundary, like a fence. It is found by adding together the lengths of all of the sides of a figure. Since a square has four equal sides, its perimeter can be calculated by multiplying the length of one side by 4. Thus, the formula is $P = 4 \times s$, where s equals one side. Like a square, a rectangle's perimeter is measured by adding together all of the sides. But as the sides are unequal, the formula is different. A rectangle has equal values for its lengths (long sides) and equal values for its widths (short sides), so the perimeter formula for a rectangle is $P = l + l + w + w = 2l + 2w$, where l is length and w is width. Perimeter is measured in simple units such as inches, feet, yards, centimeters, meters, miles, etc.

In contrast to perimeter, area is the space occupied by a defined enclosure, like a field enclosed by a fence. It is measured in square units such as square feet or square miles. Here are some formulas for the areas of basic planar shapes:

- The area of a rectangle is $l \times w$, where w is the width and l is the length
- The area of a square is s^2, where s is the length of one side (this follows from the formula for rectangles)
- The area of a triangle with base b and height h is $\frac{1}{2}bh$
- The area of a circle with radius r is πr^2

Volume is the measurement of how much space an object occupies, like how much space is in the cube. Volume questions will typically ask how much of something is needed to completely fill the object. It is measured in cubic units, such as cubic inches. Here are some formulas for the volumes of basic three-dimensional geometric figures:

- For a regular prism whose sides are all rectangles, the volume is $l \times w \times h$, where w is the width, l is the length, and h is the height of the prism

- For a cube, which is a prism whose faces are all squares of the same size, the volume is s^3

- The volume of a sphere of radius r is given by $\frac{4}{3}\pi r^3$

- The volume of a cylinder whose base has a radius of r and which has a height of h is given by $\pi r^2 h$

Word Problems

Word problems can appear daunting, but don't let the verbiage psych you out. No matter the scenario or specifics, the key to answering them is to translate the words into a math problem. Always keep in mind what the question is asking and what operations could lead to that answer.

Translating Words into Math

When asked to rewrite a mathematical expression as a situation or translated from a word problem into an expression, look for a series of key words indicating addition, subtraction, multiplication, or division:

Addition: add, altogether, together, plus, increased by, more than, in all, sum, and total

Subtraction: minus, less than, difference, decreased by, fewer than, remain, and take away

Multiplication: *times, twice, of, double,* and *triple*

Division: divided by, cut up, half, quotient of, split, and shared equally

Identifying and utilizing the proper units for the scenario requires knowing how to apply the conversion rates for money, length, volume, and mass. For example, given a scenario that requires subtracting 8 inches from $2\frac{1}{2}$ feet, both values should first be expressed in the same unit (they could be expressed $\frac{2}{3}$ft & $2\frac{1}{2}$ft, or 8in and 30in). The desired unit for the answer may also require converting back to another unit.

Consider the following scenario: A parking area along the river is only wide enough to fit one row of cars and is $\frac{1}{2}$ kilometers long. The average space needed per car is 5 meters. How many cars can be parked along the river? First, all measurements should be converted to similar units: $\frac{1}{2}$km = 500m. The operation(s) needed should be identified. Because the problem asks for the number of cars, the total space should be divided by the space per car. 500 meters divided by 5 meters per car yields a total of 100 cars. Written as an expression, the meters unit cancels and the cars unit is left: $\frac{500m}{5m/car}$ is the same as $500m \times \frac{1\ car}{5m}$ yields 100 cars.

When dealing with problems involving elapsed time, breaking the problem down into workable parts is helpful. For example, suppose the length of time between 1:15pm and 3:45pm must be determined. From 1:15pm to 2:00pm is 45 minutes (knowing there are 60 minutes in an hour). From 2:00pm to 3:00pm is 1 hour. From 3:00pm to 3:45pm is 45 minutes. The total elapsed time is 45 minutes plus 1 hour plus 45 minutes. This sum produces 1 hour and 90 minutes. 90 minutes is over an hour, so this is converted to 1 hour (60 minutes) and 30 minutes. The total elapsed time can now be expressed as 2 hours and 30 minutes.

Example 1
Alexandra made $96 during the first 3 hours of her shift as a temporary worker at a law office. She will continue to earn money at this rate until she finishes in 5 more hours. How much does Alexandra make per hour? How much will Alexandra have made at the end of the day?

The hourly rate can be figured by dividing $96 by 3 hours to get $32 per hour. Now her total pay can be figured by multiplying $32 per hour by 8 hours, which comes out to $256.

Example 2
Bernard wishes to paint a wall that measures 20 feet wide by 8 feet high. It costs $0.10 to paint 1 square foot. How much money will Bernard need for paint?

The final quantity to compute is the *cost* to paint the wall. This will be ten cents ($0.10) for each square foot of area needed to paint. The area to be painted is unknown, but the dimensions of the wall are given; thus, it can be calculated.

The dimensions of the wall are 20 feet wide and 8 feet high. Since the area of a rectangle is length multiplied by width, the area of the wall is $8 \times 20 = 160$ square feet. Multiplying 0.1 x 160 yields $16 as the cost of the paint.

Data Analysis

Representing Data
Most statistics involve collecting a large amount of data, analyzing it, and then making decisions based on previously known information. These decisions also can be measured through additional data collection and then analyzed. Therefore, the cycle can repeat itself over and over. Representing the data visually is a large part of the process, and many plots on the real number line exist that allow this to be done. For example, a *dot plot* uses dots to represent data points above the number line. Also, a *histogram* represents a data set as a collection of rectangles, which illustrate the frequency distribution of the data. Finally, a *box plot* (also known as a *box and whisker plot*) plots a data set on the number line by segmenting the distribution into four quartiles that are divided equally in half by the median.

Here's an example of a box plot, a histogram, and a dot plot for the same data set:

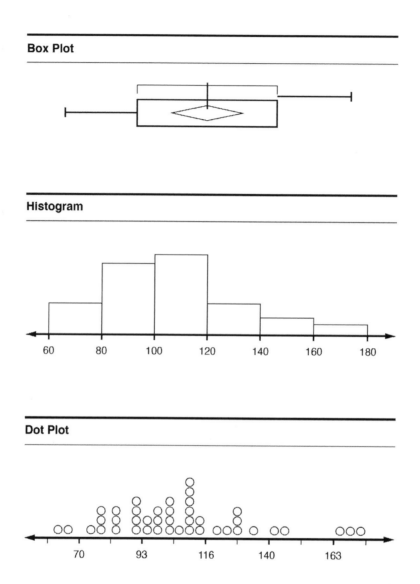

Comparing Data

Comparing data sets within statistics can mean many things. The first way to compare data sets is by looking at the center and spread of each set. The center of a data set can mean two things: median or mean. The *median* is the value that's halfway into each data set, and it splits the data into two intervals. The *mean* is the average value of the data within a set. It's calculated by adding up all of the data in the set and dividing the total by the number of data points. Outliers can significantly impact the mean. Additionally, two completely different data sets can have the same mean. For example, a data set with values ranging from 0 to 100 and a data set with values ranging from 44 to 56 can both have means of 50. The first data set has a much wider range, which is known as the *spread* of the data. This measures how varied the data is within each set. Spread can be defined further as either interquartile range or standard deviation. The *interquartile range (IQR)* is the range of the middle 50 percent of the data set. This range can be seen in the large rectangle on a box plot. The *standard deviation (σ)* quantifies the amount of variation with respect to the mean. A lower standard deviation shows that the data set doesn't differ greatly from the mean. A larger standard deviation shows that the data set is spread out farther from the mean. The formula for standard deviation is:

$$\sigma = \sqrt{\frac{\Sigma(x - \bar{x})^2}{n - 1}}$$

x is each value in the data set, $\bar{x}$ is the mean, and n is the total number of data points in the set.

<u>Interpreting Data</u>
The shape of a data set is another way to compare two or more sets of data. If a data set isn't symmetric around its mean, it's said to be *skewed.* If the tail to the left of the mean is longer, it's said to be *skewed to the left.* In this case, the mean is less than the median. Conversely, if the tail to the right of the mean is longer, it's said to be *skewed to the right* and the mean is greater than the median. When classifying a data set according to its shape, its overall *skewness* is being discussed. If the mean and median are equal, the data set isn't skewed; it is *symmetric.*

An outlier is a data point that lies a great distance away from the majority of the data set. It also can be labeled as an extreme value. Technically, an outlier is any value that falls 1.5 times the IQR above the upper quartile or 1.5 times the IQR below the lower quartile. The effect of outliers in the data set is seen visually because they affect the mean. If there's a large difference between the mean and mode, outliers are the cause. The mean shows bias toward the outlying values. However, the median won't be affected as greatly by outliers.

<u>Normal Distribution</u>
A *normal distribution* of data follows the shape of a bell curve and the data set's median, mean, and mode are equal. Therefore, 50 percent of its values are less than the mean and 50 percent are greater than the mean. Data sets that follow this shape can be generalized using normal distributions. Normal distributions are described as *frequency distributions* in which the data set is plotted as percentages rather than true data points. A *relative frequency distribution* is one where the y-axis is between zero and 1, which is the same as 0% to 100%. Within a standard deviation, 68 percent of the values are within 1 standard deviation of the mean, 95 percent of the values are within 2 standard deviations of the mean, and 99.7 percent of the values are within 3 standard deviations of the mean. The number of standard deviations that a data point falls from the mean is called the *z-score.* The formula for the z-score is $Z = \frac{x - \mu}{\sigma}$, where μ is the mean, σ is the standard deviation, and x is the data point. This formula

is used to fit any data set that resembles a normal distribution to a standard normal distribution, in a process known as *standardizing*. Here is a normal distribution with labeled z-scores:

Normal Distribution with Labelled Z-Scores

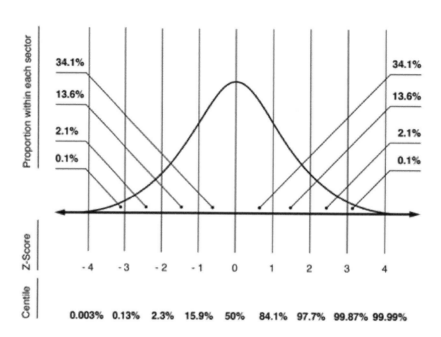

Population percentages can be estimated using normal distributions. For example, the probability that a data point will be less than the mean, or that the z-score will be less than 0, is 50%. Similarly, the probability that a data point will be within 1 standard deviation of the mean, or that the z-score will be between -1 and 1, is about 68.2%. When using a table, the left column states how many standard deviations (to one decimal place) away from the mean the point is, and the row heading states the second decimal place. The entries in the table corresponding to each column and row give the probability, which is equal to the area.

Areas Under the Curve

The area under the curve of a standard normal distribution is equal to 1. Areas under the curve can be estimated using the z-score and a table. The area is equal to the probability that a data point lies in that region in decimal form. For example, the area under the curve from $z = -1$ to $z = 1$ is 0.682.

Algebra

Computation with Integers and Negative Rational Numbers

Integers are the whole numbers together with their negatives. They include numbers like 5, 24, 0, -6, and 15. They do not include fractions or numbers that have digits after the decimal point.

Rational numbers are all numbers that can be written as a fraction using integers. A *fraction* is written as $\frac{x}{y}$ and represents the quotient of x being divided by y. More practically, it means dividing the whole into y equal parts, then taking x of those parts.

Examples of rational numbers include $\frac{1}{2}$ and $\frac{5}{4}$. The number on the top is called the *numerator*, and the number on the bottom is called the *denominator*. Because every integer can be written as a fraction with a denominator of 1, (e.g. $\frac{3}{1} = 3$), every integer is also a rational number.

When adding integers and negative rational numbers, there are some basic rules to determine if the solution is negative or positive:

Adding two positive numbers results in a positive number: 3.3 + 4.8 = 8.1.

Adding two negative numbers results in a negative number: (-8) + (-6) = -14.

Adding one positive and one negative number requires taking the absolute values and finding the difference between them. Then, the sign of the number that has the higher absolute value for the final solution is used.

For example, (-9) + 11, has a difference of absolute values of 2. The final solution is 2 because 11 has the higher absolute value. Another example is 9 + (-11), which has a difference of absolute values of 2. The final solution is -2 because 11 has the higher absolute value.

When subtracting integers and negative rational numbers, one has to change the problem to adding the opposite and then apply the rules of addition.

Subtracting two positive numbers is the same as adding one positive and one negative number.

For example, $4.9 - 7.1$ is the same as $4.9 + (-7.1)$. The solution is -2.2 since the absolute value of -7.1 is greater. Another example is $8.5 - 6.4$ which is the same as $8.5 + (-6.4)$. The solution is 2.1 since the absolute value of 8.5 is greater.

Subtracting a positive number from a negative number results in negative value.

For example, $(-12) - 7$ is the same as $(-12) + (-7)$ with a solution of -19.

Subtracting a negative number from a positive number results in a positive value.

For example, $12 - (-7)$ is the same as $12 + 7$ with a solution of 19.

For multiplication and division of integers and rational numbers, if both numbers are positive or both numbers are negative, the result is a positive value.

For example, $(-1.7)(-4)$ has a solution of 6.8 since both numbers are negative values.

If one number is positive and another number is negative, the result is a negative value.

For example, $\frac{-15}{5}$ has a solution of -3 since there is one negative number.

The Use of Absolute Values

The *absolute value* represents the distance a number is from 0. The *absolute value symbol* is | | with a number between the bars. The |10| = 10 and the |-10| = 10.

When simplifying an algebraic expression, the value of the absolute value expression is determined first, much like parenthesis in the order of operations. See the example below:

$$|8 - 12| + 5 = |-4| + 5 = 4 + 5 = 9$$

Ordering

Exponents are shorthand for longer multiplications or divisions. The exponent is written to the upper right of a number. In the expression 2^3, the exponent is 3. The number with the exponent is called the *base*.

When the exponent is a whole number, it means to multiply the base by itself as many times as the number in the exponent. So, $2^3 = 2 \times 2 \times 2 = 8$.

If the exponent is a negative number, it means to take the reciprocal of the positive exponent:

$$2^{-3} = \frac{1}{2^3} = \frac{1}{8}$$

When the exponent is 0, the result is always 1: $2^0 = 1, 5^0 = 1$, and so on.

When the exponent is 2, the number is *squared*, and when the exponent is 3, it is *cubed*.

When working with longer expressions, parentheses are used to show the order in which the operations should be performed. Operations inside the parentheses should be completed first. Thus, $(3 - 1) \div 2$ means one should first subtract 1 from 3, and then divide that result by 2.

The *order of operations* gives an order for how a mathematical expression is to be simplified:

- Parentheses
- Exponents
- Multiplication
- Division
- Addition
- Subtraction

To help remember this, many students like to use the mnemonic PEMDAS. Some students associate this word with a phrase to help them, such as "Pirates Eat Many Donuts at Sea." Here is a quick example:

Evaluate $2^2 \times (3 - 1) \div 2 + 3$.

Parenthesis: $2^2 \times 2 \div 2 + 3$.

Exponents: $4 \times 2 \div 2 + 3$

Multiply: $8 \div 2 + 3$.

Divide: $4 + 3$.

Addition: 7

Evaluation of Simple Formulas and Expressions

To evaluate simple formulas and expressions, the first step is to substitute the given values in for the variable(s). Then, the order of operations is used to simplify.

Example 1
Evaluate $\frac{1}{2}x^2 - 3, x = 4$.

The first step is to substitute in 4 for x in the expression: $\frac{1}{2}(4)^2 - 3$.

Then, the order of operations is used to simplify.

The exponent comes first, $\frac{1}{2}(16) - 3$, then the multiplication $8 - 3$, and then, after subtraction, the solution is 5.

Example 2
Evaluate $4|5 - x| + 2y, x = 4, y = -3$.

The first step is to substitute 4 in for x and -3 in for y in the expression: $4|5 - 4| + 2(-3)$.

Then, the absolute value expression is simplified, which is $|5 - 4| = |1| = 1$.

The expression is $4(1) + 2(-3)$ which can be simplified using the order of operations.

First is the multiplication, $4 + (-6)$; then addition yields an answer of -2.

Example 3
Find the perimeter of a rectangle with a length of 6 inches and a width of 9 inches.

The first step is substituting in 6 for the length and 9 for the width in the perimeter of a rectangle formula, $P = 2(6) + 2(9)$.

Then, the order of operations is used to simplify.

First is multiplication (resulting in $12 + 18$) and then addition for a solution of 30 inches.

Adding and Subtracting Monomials and Polynomials

To add or subtract polynomials, add the coefficients of terms with the same exponent. For instance,

$$(-2x^2 + 3x + 1) + (4x^2 - x)$$

$$(-2 + 4)x^2 + (3 - 1)x + 1$$

$$2x^2 + 2x + 1$$

Multiplying and Dividing Monomials and Polynomials

To multiply polynomials, each term of the first polynomial multiplies each term of the second polynomial and adds up the results. Here's an example:

$$(3x^4 + 2x^2)(2x^2 + 3)$$

$$3x^4 \times 2x^2 + 3x^4 \times 3 + 2x^2 \times 2x^2 + 2x^2 \times 3$$

Then, add like terms with a result of:

$$6x^6 + 9x^4 + 4x^4 + 6x^2 = 6x^6 + 13x^4 + 6x^2$$

A polynomial with two terms is called a *binomial*. Another way to remember the rule for multiplying two binomials is to use the acronym *FOIL*: multiply the *First* terms together, then the *Outside* terms (terms on the far left and far right), then the *Inner* terms, and finally the *Last* two terms. For longer polynomials, there is no such convenient mnemonic, so remember to multiply each term of the first polynomial by each term of the second, and add the results.

To divide one polynomial by another, the procedure is similar to long division. At each step, one needs to figure out how to get the term of the dividend with the highest exponent as a multiple of the divisor. The divisor is multiplied by the multiple to get that term, which goes in the quotient. Then, the product of this term is subtracted with the dividend from the divisor and repeat the process. An example of polynomial long division is shown below.

Example
$(4x^3 + x^2 - x + 4) \div (2x - 1)$

The long division is carried out as follows:

$$
\begin{array}{r}
2x^2 + \dfrac{3}{2}x + \dfrac{1}{4} \\[2mm]
\hline
2x - 1 \,\big)\, 4x^3 + x^2 - x + 4 \\[1mm]
-\ \ 4x^3 - 2x^2 \\[1mm]
\hline
3x^2 - x \\[1mm]
-\ \ \ \ 3x^2 - \dfrac{3}{2}x \\[1mm]
\hline
\dfrac{1}{2}x + 4 \\[1mm]
-\ \ \ \ \dfrac{1}{2}x - \dfrac{1}{4} \\[1mm]
\hline
\dfrac{17}{4}
\end{array}
$$

The remainder must be placed over the original divisor and the fraction removed from the numerator as follows:

$$\frac{\frac{17}{4}}{2x - 1} = \frac{17}{8x - 4}$$

The answer is:

$$2x^2 + \frac{3}{2}x + \frac{1}{4} + \frac{17}{8x - 4}$$

To get $4x^3$ from the second polynomial, multiply by $2x^2$.

The first term for the quotient is going to be $2x^2$.

The result of $2x^2(2x - 1)$ is $4x^3 - 2x^2$. Subtract this from the first polynomial.

The result is:

$$(3x^2 - x + 4) \div (2x - 1)$$

The procedure is repeated: to cancel the $-x^2$ term, then multiply $(2x - 1)$ by $-\frac{1}{2}x$.

Adding this to the quotient, the quotient becomes $2x^2 - \frac{1}{2}x$.

The dividend is changed by subtracting $-\frac{1}{2}x(2x - 1)$ from it to obtain:

$$(-\frac{3}{2}x + 4) \div (2x - 1)$$

To get $-\frac{3}{2}x$ needs to be multiplied by $-\frac{3}{4}$.

The quotient, therefore, becomes:

$$2x^2 - \frac{1}{2}x - \frac{3}{4}$$

The remaining part is:

$$4.75 \div (2x - 1)$$

There is no monomial to multiply to cancel this constant term, since the divisor now has a higher power than the dividend.

The final answer is the quotient plus the remainder divided by $(2x - 1)$:

$$2x^2 - \frac{1}{2}x - \frac{3}{4} + \frac{4.75}{2x - 1}$$

The Evaluation of Positive Rational Roots and Exponents

There are a few rules for working with exponents. For any numbers a, b, m, n, the following hold true:

$$a^1 = a$$

$$1^a = 1$$

$$a^0 = 1$$

$$a^m \times a^n = a^{m+n}$$

$$a^m \div a^n = a^{m-n}$$

$$(a^m)^n = a^{m \times n}$$

$$(a \times b)^m = a^m \times b^m$$

$$(a \div b)^m = a^m \div b^m$$

Any number, including a fraction, can be an exponent. The same rules apply.

Simplifying Algebraic Fractions

A *rational expression* is a fraction with a polynomial in the numerator and denominator. The denominator polynomial cannot be zero. An example of a rational expression is $\frac{3x^4 - 2}{-x+1}$. The same rules for working with addition, subtraction, multiplication, and division with rational expressions apply as when working with regular fractions.

The first step is to find a common denominator when adding or subtracting. This can be done just as with regular fractions. For example, if $\frac{a}{b} + \frac{c}{d}$, then a common denominator can be found by multiplying to find the following fractions: $\frac{ad}{bd}, \frac{cb}{db}$.

A *complex fraction* is a fraction in which the numerator and denominator are themselves fractions, of the form $\frac{\left(\frac{a}{b}\right)}{\left(\frac{c}{d}\right)}$. These can be simplified by following the usual rules for the order of operations, or by remembering that dividing one fraction by another is the same as multiplying by the reciprocal of the divisor. This means that any complex fraction can be rewritten using the following form:

$$\frac{\left(\frac{a}{b}\right)}{\left(\frac{c}{d}\right)} = \frac{a}{b} \times \frac{d}{c}$$

The following problem is an example of solving a complex fraction:

$$\frac{\left(\frac{5}{4}\right)}{\left(\frac{3}{8}\right)} = \frac{5}{4} \times \frac{8}{3} = \frac{40}{12} = \frac{10}{3}$$

Factoring

Factors for polynomials are similar to factors for integers. One polynomial is a factor of a second polynomial if the second polynomial can be obtained from the first by multiplying by a third polynomial. $6x^6 + 13x^4 + 6x^2$ can be obtained by multiplying $(3x^4 + 2x^2)$ and $(2x^2 + 3)$ together. This means $2x^2 + 3$ and $3x^4 + 2x^2$ are factors of $6x^6 + 13x^4 + 6x^2$.

In general, finding the factors of a polynomial can be tricky. However, there are a few types of polynomials that can be factored in a straightforward way. If a certain monomial divides each term of a polynomial, it can be factored out:

$$x^2 + 2xy + y^2 = (x + y)^2$$

$$x^2 - 2xy + y^2 = (x - y)^2$$

$$x^2 - y^2 = (x + y)(x - y)$$

$$x^3 + y^3 = (x + y)(x^2 - xy + y^2)$$

$$x^3 - y^3 = (x - y)(x^2 + xy + y^2)$$

$$x^3 + 3x^2y + 3xy^2 + y^3 = (x + y)^3$$

$$x^3 - 3x^2y + 3xy^2 - y^3 = (x - y)^3$$

These rules can be used in many combinations with one another. To give one example, the expression $3x^3 - 24$ factors to

$$3(x^3 - 8)$$

$$3(x - 2)(x^2 + 2x + 4)$$

When factoring polynomials, it is a good idea to multiply the factors to check the result.

Solving Linear Equations and Inequalities

The simplest equations to solve are *linear equations*, which have the form $ax + b = 0$. These have the solution $x = -\frac{b}{a}$.

For instance, in the equation $\frac{1}{3}x - 4 = 0$, it can be determined that $\frac{1}{3}x = 4$ by adding 4 on each side. Next, both sides of the equation are multiplied by 3 to get $x = 12$.

Solving an inequality is very similar to solving equations, with one important issue to keep track of: if multiplying or dividing both sides of an inequality by a negative number, the direction of the inequality *flips*.

35

For example, consider the inequality $-4x < 12$. Solving this inequality requires the division of -4. When the negative four is divided, the less-than sign changes to a greater-than sign. The solution becomes $x > -3$.

Example
$-4x - 3 \leq -2x + 1$

2x is added to both sides, and 3 is added to both sides, leaving $-2x \leq 4$.

$-2x \leq 4$ is multiplied by $-\frac{1}{2}$, which means flipping the direction of the inequality.

This gives $x \geq -2$.

An *absolute inequality* is an inequality that is true for all real numbers. An inequality that is only true for some real numbers is called *conditional*.

In addition to the inequalities above, there are also *double inequalities* where three quantities are compared to one another, such as $3 \leq x + 4 < 5$. The rules for double inequalities include always performing any operations to every part of the inequality and reversing the direction of the inequality when multiplying or dividing by a negative number.

When solving equations and inequalities, the solutions can be checked by plugging the answer back in to the original problem. If the solution makes a true statement, the solution is correct.

Solving Quadratic Equations by Factoring

Solving quadratic equations is a little trickier. If they take the form $ax^2 - b = 0$, then the equation can be solved by adding b to both sides and dividing by a to get $x^2 = \frac{b}{a}$.

Using the sixth rule above, the solution is $x = \pm\sqrt{\frac{b}{a}}$. Note that this is actually two separate solutions, unless b happens to be 0.

If a quadratic equation has no constant—so that it takes the form $ax^2 + bx = 0$—then the x can be factored out to get $x(ax + b) = 0$. Then, the solutions are $x = 0$, together with the solutions to $ax + b = 0$. Both factors x and $(ax + b)$ can be set equal to zero to solve for x because one of those values must be zero for their product to equal zero. For an equation $ab = 0$ to be true, either $a = 0$, or $b = 0$.

A given quadratic equation $x^2 + bx + c$ can be factored into $(x + A)(x + B)$, where $A + B = b$, and $AB = c$. Finding the values of A and B can take time, but such a pair of numbers can be found by guessing and checking. Looking at the positive and negative factors for c offers a good starting point.

For example, in $x^2 - 5x + 6$, the factors of 6 are 1, 2, and 3. Now, $(-2)(-3) = 6$, and $-2 - 3 = -5$. In general, however, this may not work, in which case another approach may need to be used.

A quadratic equation of the form $x^2 + 2xb + b^2 = 0$ can be factored into $(x + b)^2 = 0$. Similarly, $x^2 - 2xy + y^2 = 0$ factors into $(x - y)^2 = 0$.

In general, the constant term may not be the right value to be factored this way. A more general method for solving these quadratic equations must then be found. The following two methods will work in any situation.

36

Completing the Square

The first method is called *completing the square*. The idea here is that in any equation $x^2 + 2xb + c = 0$, something could be added to both sides of the equation to get the left side to look like $x^2 + 2xb + b^2$, meaning it could be factored into $(x + b)^2 = 0$.

Example
$x^2 + 6x - 1 = 0$

The left-hand side could be factored if the constant were equal to 9, since $x^2 + 6x + 9 = (x + 3)^2$.

To get a constant of 9 on the left, 10 needs to be added to both sides.

That changes the equation to $x^2 + 6x + 9 = 10$.

Factoring the left gives $(x + 3)^2 = 10$.

Then, the square root of both sides can be taken (remembering that this introduces a $\pm$): $x + 3 = \pm\sqrt{10}$.

Finally, 3 is subtracted from both sides to get two solutions: $x = -3 \pm \sqrt{10}$.

The Quadratic Formula

The first method of completing the square can be used in finding the second method, the quadratic formula. It can be used to solve any quadratic equation. This formula may be the longest method for solving quadratic equations and is commonly used as a last resort after other methods are ruled out.

It can be helpful in memorizing the formula to see where it comes from, so here are the steps involved.

The most general form for a quadratic equation is $ax^2 + bx + c = 0$.

First, dividing both sides by a leaves us with $x^2 + \frac{b}{a}x + \frac{c}{a} = 0$.

To complete the square on the left-hand side, $\frac{c}{a}$ can be subtracted on both sides to get $x^2 + \frac{b}{a}x = -\frac{c}{a}$.

$(\frac{b}{2a})^2$ is then added to both sides.

This gives $x^2 + \frac{b}{a}x + (\frac{b}{2a})^2 = (\frac{b}{2a})^2 - \frac{c}{a}$.

The left can now be factored and the right-hand side simplified to give $(x + \frac{b}{2a})^2 = \frac{b^2 - 4ac}{4a}$.

Taking the square roots gives:

$$x + \frac{b}{2a} = \pm\frac{\sqrt{b^2 - 4ac}}{2a}$$

Solving for x yields the quadratic formula:

$$x = \frac{-b \pm \sqrt{b^2 - 4ac}}{2a}$$

It isn't necessary to remember how to get this formula but memorizing the formula itself is the goal.

If an equation involves taking a root, then the first step is to move the root to one side of the equation and everything else to the other side. That way, both sides can be raised to the index of the radical in order to remove it, and solving the equation can continue.

Solving Verbal Problems Presented in an Algebraic Context

There is a four-step process in problem-solving that can be used as a guide:

- Understand the problem and determine the unknown information.
- Translate the verbal problem into an algebraic equation.
- Solve the equation by using inverse operations.
- Check the work and answer the given question.

Example
Three times the sum of a number plus 4 equals the number plus 8. What is the number?

The first step is to determine the unknown, which is the number, or x.

The second step is to translate the problem into the equation, which is $3(x + 4) = x + 8$.

The equation can be solved as follows:

$3x + 12 = x + 8$	Apply the distributive property
$3x = x - 4$	Subtract 12 from both sides of the equation
$2x = -4$	Subtract x from both sides of the equation
$x = -2$	Divide both sides of the equation by 2

The final step is checking the solution. Plugging the value for x back into the equation yields the following problem: $3(-2) + 12 = -2 + 8$. Using the order of operations shows that a true statement is made: $6 = 6$.

Geometry

Simple Geometry Problems

There are many key facts related to geometry that are applicable. The sum of the measures of the angles of a triangle are 180°, and for a quadrilateral, the sum is 360°. Rectangles and squares each have four right angles. A *right angle* has a measure of 90°.

Perimeter

The *perimeter* is the distance around a figure or the sum of all sides of a polygon.

The *formula for the perimeter of a square* is four times the length of a side. For example, the following square has side lengths of 5 feet:

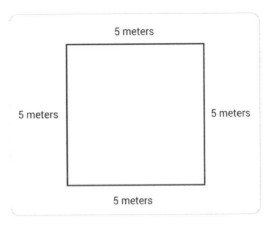

The perimeter is 20 feet because 4 times 5 is 20.

The *formula for a perimeter of a rectangle* is the sum of twice the length and twice the width. For example, if the length of a rectangle is 10 inches and the width 8 inches, then the perimeter is 36 inches because $P = 2l + 2w = 2(10) + 2(8) = 20 + 16 = 36$ inches.

Area

The area is the amount of space inside of a figure, and there are formulas associated with area.

The area of a triangle is the product of one-half the base and height. For example, if the base of the triangle is 2 feet and the height 4 feet, then the area is 4 square feet. The following equation shows the formula used to calculate the area of the triangle:

$$A = \frac{1}{2}bh = \frac{1}{2}(2)(4) = 4 \text{ square feet}$$

The area of a square is the length of a side squared, and the area of a rectangle is length multiplied by the width. For example, if the length of the square is 7 centimeters, then the area is 49 square centimeters. The formula for this example is $A = s^2 = 7^2 = 49$ square centimeters. An example is if the rectangle has a length of 6 inches and a width of 7 inches, then the area is 42 square inches:

$$A = lw = 6(7) = 42 \text{ square inches}$$

The area of a trapezoid is one-half the height times the sum of the bases. For example, if the length of the bases are 2.5 and 3 feet and the height 3.5 feet, then the area is 9.625 square feet. The following formula shows how the area is calculated:

$$A = \frac{1}{2}h(b_1 + b_2)$$

$$\frac{1}{2}(3.5)(2.5 + 3)$$

$$\frac{1}{2}(3.5)(5.5) = 9.625 \text{ square feet}$$

The perimeter of a figure is measured in single units, while the area is measured in square units.

Distribution of a Quantity into its Fractional Parts

A quantity may be broken into its fractional parts. For example, a toy box holds three types of toys for kids. $\frac{1}{3}$ of the toys are Type A and $\frac{1}{4}$ of the toys are Type B. With that information, how many Type C toys are there?

First, the sum of Type A and Type B must be determined by finding a common denominator to add the fractions. The lowest common multiple is 12, so that is what will be used. The sum is:

$$\frac{1}{3} + \frac{1}{4} = \frac{4}{12} + \frac{3}{12} = \frac{7}{12}$$

This value is subtracted from 1 to find the number of Type C toys. The value is subtracted from 1 because 1 represents a whole. The calculation is:

$$1 - \frac{7}{12}$$

$$\frac{12}{12} - \frac{7}{12} = \frac{5}{12}$$

This means that $\frac{5}{12}$ of the toys are Type C. To check the answer, add all fractions together, and the result should be 1.

Plane Geometry

Locations on the plane that have no width or breadth are called *points*. These points usually will be denoted with capital letters such as *P*.

Any pair of points *A*, *B* on the plane will determine a unique straight line between them. This line is denoted *AB*. Sometimes to emphasize a line is being considered, this will be written as $\overleftrightarrow{AB}$.

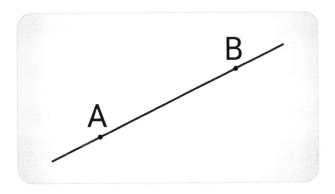

If the Cartesian coordinates for *A* and *B* are known, then the distance $d(A, B)$ along the line between them can be measured using the *Pythagorean formula*, which states that if $A = (x_1, y_1)$ and $B = (x_2, y_2)$, then the distance between them is:

$$d(A, B) = \sqrt{(x_2 - x_1)^2 + (y_2 - y_1)^2}$$

The part of a line that lies between *A* and *B* is called a *line segment*. It has two endpoints, one at *A* and one at *B*. *Rays* also can be formed. Given points *A* and *B*, a *ray* is the portion of a line that starts at one of these points, passes through the other, and keeps on going. Therefore, a ray has a single endpoint, but the other end goes off to infinity.

Given a pair of points *A* and *B*, a circle centered at *A* and passing through *B* can be formed. This is the set of points whose distance from *A* is exactly $d(A, B)$. The radius of this circle will be $d(A, B)$.

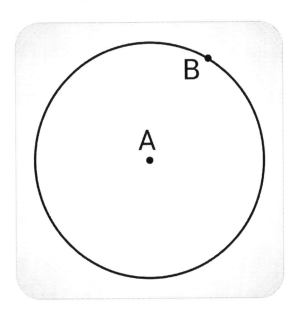

The *circumference* of a circle is the distance traveled by following the edge of the circle for one complete revolution, and the length of the circumference is given by $2\pi r$, where r is the radius of the circle. The formula for circumference is $C = 2\pi r$.

When two lines cross, they form an *angle*. The point where the lines cross is called the *vertex* of the angle. The angle can be named by either just using the vertex, $\angle A$, or else by listing three points $\angle BAC$, as shown in the diagram below.

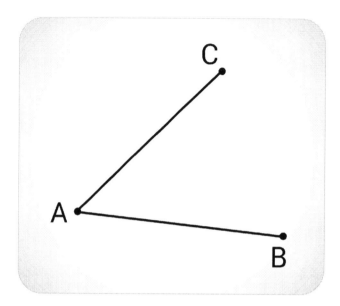

The measurement of an angle can be given in degrees or in radians. In degrees, a full circle is 360 degrees, written 360°. In radians, a full circle is 2π radians.

Given two points on the circumference of a circle, the path along the circle between those points is called an *arc* of the circle. For example, the arc between *B* and *C* is denoted by a thinner line:

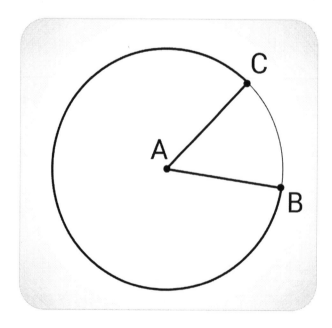

The length of the path along an arc is called the *arc length*. If the circle has radius r, then the arc length is given by multiplying the measure of the angle in radians by the radius of the circle.

Two lines are said to be *parallel* if they never intersect. If the lines are *AB* and *CD*, then this is written as $AB \parallel CD$.

If two lines cross to form four quarter-circles, that is, 90° angles, the two lines are *perpendicular*. If the point at which they cross is *B*, and the two lines are *AB* and *BC*, then this is written as $AB \perp BC$.

A *polygon* is a closed figure (meaning it divides the plane into an inside and an outside) consisting of a collection of line segments between points. These points are called the *vertices* of the polygon. These line segments must not overlap one another. Note that the number of sides is equal to the number of angles, or vertices of the polygon. The angles between line segments meeting one another in the polygon are called *interior angles*.

A *regular polygon* is a polygon whose edges are all the same length and whose interior angles are all of equal measure.

A *triangle* is a polygon with three sides. A *quadrilateral* is a polygon with four sides.

A *right triangle* is a triangle that has one 90° angle.

The sum of the interior angles of any triangle must add up to 180°.

An *isosceles triangle* is a triangle in which two of the sides are the same length. In this case, it will always have two congruent interior angles. If a triangle has two congruent interior angles, it will always be isosceles.

An *equilateral triangle* is a triangle whose sides are all the same length and whose angles are all equivalent to one another, equal to 60°. Equilateral triangles are examples of regular polygons. Note that equilateral triangles are also isosceles.

A *rectangle* is a quadrilateral whose interior angles are all 90°. A rectangle has two sets of sides that are equal to one another.

A *square* is a rectangle whose width and height are equal. Therefore, squares are regular polygons.

A *parallelogram* is a quadrilateral in which the opposite sides are parallel and equivalent to each other.

Transformations of a Plane

Given a figure drawn on a plane, many changes can be made to that figure, including *rotation*, *translation*, and *reflection*. Rotations turn the figure about a point, translations slide the figure, and reflections flip the figure over a specified line. When performing these transformations, the original figure is called the *pre-image*, and the figure after transformation is called the *image*.

More specifically, *translation* means that all points in the figure are moved in the same direction by the same distance. In other words, the figure is slid in some fixed direction. Of course, while the entire figure is slid by the same distance, this does not change any of the measurements of the figures involved. The result will have the same distances and angles as the original figure.

In terms of Cartesian coordinates, a translation means a shift of each of the original points (x, y) by a fixed amount in the x and y directions, to become $(x + a, y + b)$.

Another procedure that can be performed is called *reflection*. To do this, a line in the plane is specified, called the *line of reflection*. Then, take each point and flip it over the line so that it is the same distance from the line but on the opposite side of it. This does not change any of the distances or angles involved, but it does reverse the order in which everything appears.

To reflect something over the x-axis, the points (x, y) are sent to $(x, -y)$. To reflect something over the y-axis, the points (x, y) are sent to the points $(-x, y)$. Flipping over other lines is not something easy to express in Cartesian coordinates. However, by drawing the figure and the line of reflection, the distance to the line and the original points can be used to find the reflected figure.

Example: Reflect this triangle with vertices (-1, 0), (2, 1), and (2, 0) over the y-axis. The pre-image is shown below.

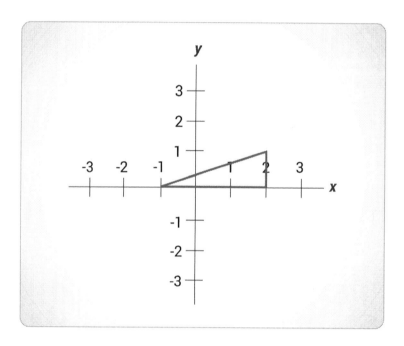

To do this, flip the x values of the points involved to the negatives of themselves, while keeping the y values the same. The image is shown here.

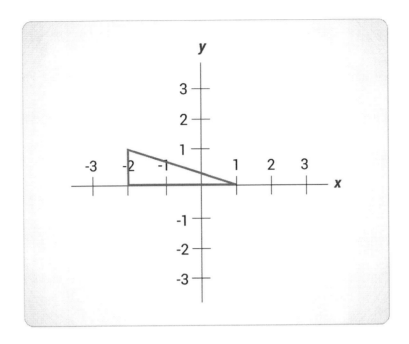

The new vertices will be (1, 0), (-2, 1), and (-2, 0).

Another procedure that does not change the distances and angles in a figure is *rotation*. In this procedure, pick a center point, then rotate every vertex along a circle around that point by the same angle. This procedure is also not easy to express in Cartesian coordinates, and this is not a requirement on this test. However, as with reflections, it's helpful to draw the figures and see what the result of the rotation would look like. This transformation can be performed using a compass and protractor.

Each one of these transformations can be performed on the coordinate plane without changes to the original dimensions or angles.

If two figures in the plane involve the same distances and angles, they are called *congruent figures*. In other words, two figures are congruent when they go from one form to another through reflection, rotation, and translation, or a combination of these.

Remember that rotation and translation will give back a new figure that is identical to the original figure, but reflection will give back a mirror image of it.

To recognize that a figure has undergone a rotation, check to see that the figure has not been changed into a mirror image, but that its orientation has changed (that is, whether the parts of the figure now form different angles with the x and y axes).

To recognize that a figure has undergone a translation, check to see that the figure has not been changed into a mirror image, and that the orientation remains the same.

To recognize that a figure has undergone a reflection, check to see that the new figure is a mirror image of the old figure.

Keep in mind that sometimes a combination of translations, reflections, and rotations may be performed on a figure.

Dilation

A *dilation* is a transformation that preserves angles, but not distances. This can be thought of as stretching or shrinking a figure. If a dilation makes figures larger, it is called an *enlargement*. If a dilation makes figures smaller, it is called a *reduction*. The easiest example is to dilate around the origin. In this case, multiply the x and y coordinates by a *scale factor*, k, sending points (x, y) to (kx, ky).

As an example, draw a dilation of the following triangle, whose vertices will be the points (-1, 0), (1, 0), and (1, 1).

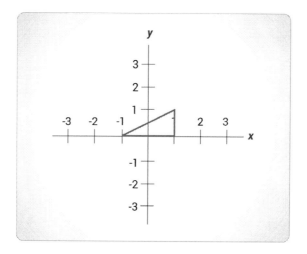

For this problem, dilate by a scale factor of 2, so the new vertices will be (-2, 0), (2, 0), and (2, 2).

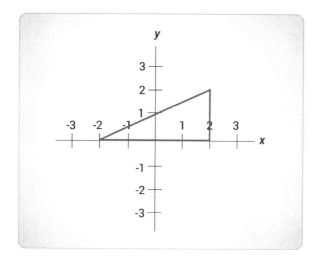

Note that after a dilation, the distances between the vertices of the figure will have changed, but the angles remain the same. The two figures that are obtained by dilation, along with possibly translation, rotation, and reflection, are all *similar* to one another. Another way to think of this is that similar figures have the same number of vertices and edges, and their angles are all the same. Similar figures have the same basic shape, but are different in size.

Symmetry

Using the types of transformations above, if an object can undergo these changes and not appear to have changed, then the figure is symmetrical. If an object can be split in half by a line and flipped over that line to lie directly on top of itself, it is said to have *line symmetry*. An example of both types of figures is seen below.

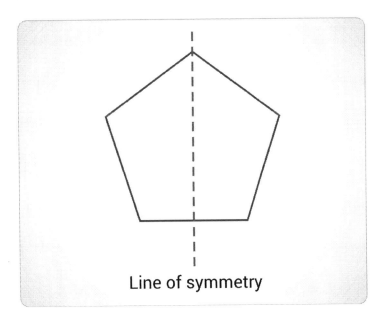

Line of symmetry

If an object can be rotated about its center to any degree smaller than 360, and it lies directly on top of itself, the object is said to have *rotational symmetry*. An example of this type of symmetry is shown below. The pentagon has an order of 5.

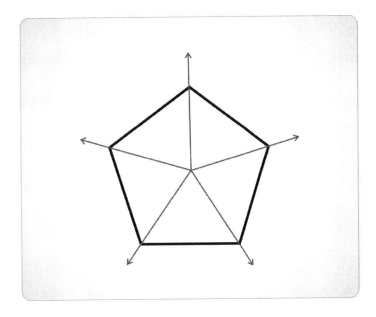

The rotational symmetry lines in the figure above can be used to find the angles formed at the center of the pentagon. Knowing that all of the angles together form a full circle, at 360 degrees, the figure can be split into 5 angles equally. By dividing the 360° by 5, each angle is 72°.

Given the length of one side of the figure, the perimeter of the pentagon can also be found using rotational symmetry. If one side length was 3 cm, that side length can be rotated onto each other side length four times. This would give a total of 5 side lengths equal to 3 cm. To find the perimeter, or distance around the figure, multiply 3 by 5. The perimeter of the figure would be 15 cm.

If a line cannot be drawn anywhere on the object to flip the figure onto itself or rotated less than or equal to 180 degrees to lay on top of itself, the object is asymmetrical. Examples of these types of figures are shown below.

No line of symmetry

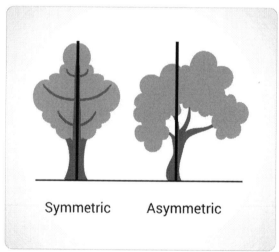

Symmetric   Asymmetric

Perimeters and Areas

The *perimeter* of a polygon is the total length of a trip around the whole polygon, starting and ending at the same point. It is found by adding up the lengths of each line segment in the polygon. For a rectangle with sides of length x and y, the perimeter will be $2x + 2y$.

The area of a polygon is the area of the region that it encloses. Regarding the area of a rectangle with sides of length x and y, the area is given by xy. For a triangle with a base of length b and a height of length h, the area is $\frac{1}{2}bh$.

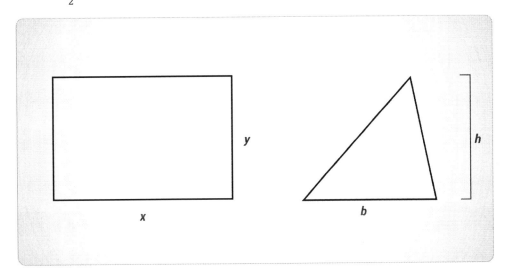

To find the areas of more general polygons, it is usually easiest to break up the polygon into rectangles and triangles. For example, find the area of the following figure whose vertices are (-1, 0), (-1, 2), (1, 3), and (1, 0).

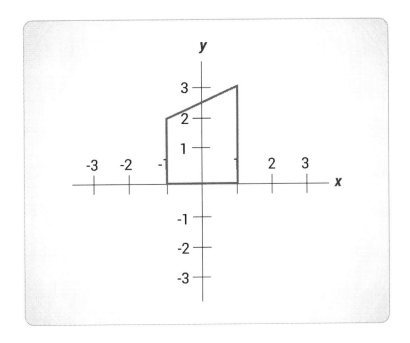

Separate this into a rectangle and a triangle as shown:

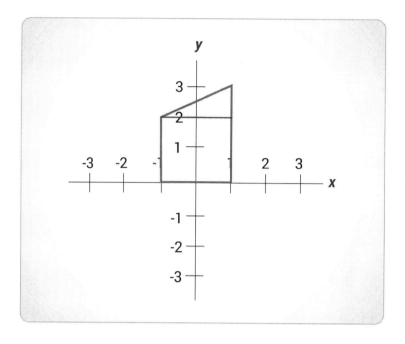

The rectangle has a height of 2 and a width of 2, so it has a total area of $2 \times 2 = 4$. The triangle has a width of 2 and a height of 1, so it has an area of $\frac{1}{2} 2 \times 1 = 1$. Therefore, the entire quadrilateral has an area of $4 + 1 = 5$.

As another example, suppose someone wants to tile a rectangular room that is 10 feet by 6 feet using triangular tiles that are 12 inches by 6 inches. How many tiles would be needed? To figure this, first find the area of the room, which will be $10 \times 6 = 60$ square feet. The dimensions of the triangle are 1 foot by ½ foot, so the area of each triangle is $\frac{1}{2} \times 1 \times \frac{1}{2} = \frac{1}{4}$ square feet. Notice that the dimensions of the triangle had to be converted to the same units as the rectangle. Now, take the total area divided by the area of one tile to find the answer: $\frac{60}{\frac{1}{4}} = 60 \times 4 = 240$ tiles required.

Volumes and Surface Areas

Geometry in three dimensions is similar to geometry in two dimensions. The main new feature is that three points now define a unique *plane* that passes through each of them. Three dimensional objects can be made by putting together two dimensional figures in different surfaces. Below, some of the possible three dimensional figures will be provided, along with formulas for their volumes and surface areas.

A rectangular prism is a box whose sides are all rectangles meeting at 90° angles. Such a box has three dimensions: length, width, and height. If the length is x, the width is y, and the height is z, then the volume is given by $V = xyz$.

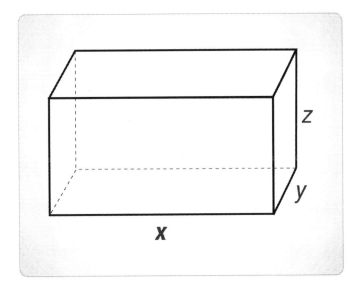

The surface area will be given by computing the surface area of each rectangle and adding them together. There are a total of six rectangles. Two of them have sides of length x and y, two have sides of length y and z, and two have sides of length x and z. Therefore, the total surface area will be given by $SA = 2xy + 2yz + 2xz$.

A *rectangular pyramid* is a figure with a rectangular base and four triangular sides that meet at a single vertex. If the rectangle has sides of length x and y, then the volume will be given by $V = \frac{1}{3}xyh$.

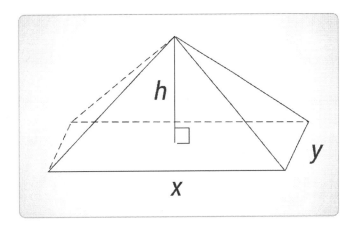

To find the surface area, the dimensions of each triangle need to be known. However, these dimensions can differ depending on the problem in question. Therefore, there is no general formula for calculating total surface area.

A *sphere* is a set of points all of which are equidistant from some central point. It is like a circle, but in three dimensions. The volume of a sphere of radius r is given by $V = \frac{4}{3}\pi r^3$. The surface area is given by $A = 4\pi r^2$.

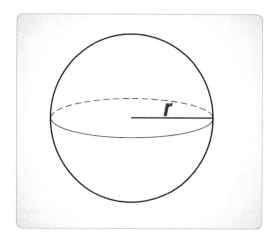

The Pythagorean Theorem

The Pythagorean theorem is an important result in geometry. It states that for right triangles, the sum of the squares of the two shorter sides will be equal to the square of the longest side (also called the *hypotenuse*). The longest side will always be the side opposite to the 90° angle. If this side is called c, and the other two sides are a and b, then the Pythagorean theorem states that $c^2 = a^2 + b^2$. Since lengths are always positive, this also can be written as $c = \sqrt{a^2 + b^2}$. A diagram to show the parts of a triangle using the Pythagorean theorem is below.

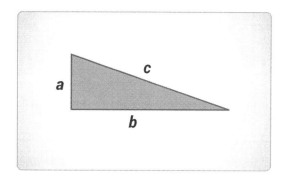

As an example of the theorem, suppose that Shirley has a rectangular field that is 5 feet wide and 12 feet long, and she wants to split it in half using a fence that goes from one corner to the opposite corner. How long will this fence need to be? To figure this out, note that this makes the field into two right triangles, whose hypotenuse will be the fence dividing it in half. Therefore, the fence length will be given by:

$$\sqrt{5^2 + 12^2} = \sqrt{169} = 13 \text{ feet long}$$

Similar Figures and Proportions

Sometimes, two figures are similar, meaning they have the same basic shape and the same interior angles, but they have different dimensions. If the ratio of two corresponding sides is known, then that ratio, or scale factor, holds true for all of the dimensions of the new figure.

Here is an example of applying this principle. Suppose that Lara is 5 feet tall and is standing 30 feet from the base of a light pole, and her shadow is 6 feet long. How high is the light on the pole? To figure this, it helps to make a sketch of the situation:

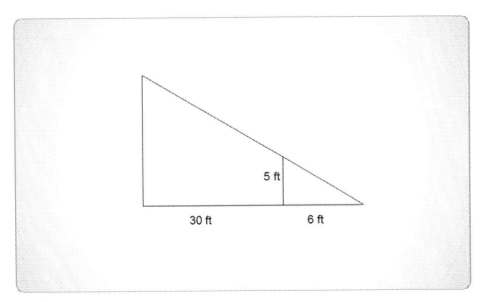

The light pole is the left side of the triangle. Lara is the 5-foot vertical line. Notice that there are two right triangles here, and that they have all the same angles as one another. Therefore, they form similar triangles. So, figure the ratio of proportionality between them.

The bases of these triangles are known. The small triangle, formed by Lara and her shadow, has a base of 6 feet. The large triangle, formed by the light pole along with the line from the base of the pole out to the end of Lara's shadow is $30 + 6 = 36$ feet long. So, the ratio of the big triangle to the little triangle will be $\frac{36}{6} = 6$. The height of the little triangle is 5 feet. Therefore, the height of the big triangle will be $6 \times 5 = 30$ feet, meaning that the light is 30 feet up the pole.

Notice that the perimeter of a figure changes by the ratio of proportionality between two similar figures, but the area changes by the *square* of the ratio. This is because if the length of one side is doubled, the area is quadrupled.

As an example, suppose two rectangles are similar, but the edges of the second rectangle are three times longer than the edges of the first rectangle. The area of the first rectangle is 10 square inches. How much more area does the second rectangle have than the first?

To answer this, note that the area of the second rectangle is $3^2 = 9$ times the area of the first rectangle, which is 10 square inches. Therefore, the area of the second rectangle is going to be $9 \times 10 = 90$ square inches. This means it has $90 - 10 = 80$ square inches more area than the first rectangle.

As a second example, suppose X and Y are similar right triangles. The hypotenuse of X is 4 inches. The area of Y is $\frac{1}{4}$ the area of X. What is the hypotenuse of Y?

First, realize the area has changed by a factor of $\frac{1}{4}$. The area changes by a factor that is the *square* of the ratio of changes in lengths, so the ratio of the lengths is the square root of the ratio of areas. That means that the ratio of lengths must be $\sqrt{\frac{1}{4}} = \frac{1}{2}$, and the hypotenuse of Y must be $\frac{1}{2} \times 4 = 2$ inches.

Volumes between similar solids change like the cube of the change in the lengths of their edges. Likewise, if the ratio of the volumes between similar solids is known, the ratio between their lengths is known by finding the cube root of the ratio of their volumes.

For example, suppose there are two similar rectangular pyramids X and Y. The base of X is 1 inch by 2 inches, and the volume of X is 8 inches. The volume of Y is 64 inches. What are the dimensions of the base of Y?

To answer this, first find the ratio of the volume of Y to the volume of X. This will be given by $\frac{64}{8} = 8$. Now the ratio of lengths is the cube root of the ratio of volumes, or $\sqrt[3]{8} = 2$. So, the dimensions of the base of Y must be 2 inches by 4 inches.

Practice Questions

1. $\dfrac{14}{15} + \dfrac{3}{5} - \dfrac{1}{30} =$

    a. $\dfrac{19}{15}$

    b. $\dfrac{43}{30}$

    c. $\dfrac{4}{3}$

    d. $\dfrac{3}{2}$

2. Solve for x and y, given $3x + 2y = 8, -x + 3y = 1$.

    a. $x = 2, y = 1$
    b. $x = 1, y = 2$
    c. $x = -1, y = 6$
    d. $x = 3, y = 1$

3. $\dfrac{1}{2}\sqrt{16} =$

    a. 0
    b. 1
    c. 2
    d. 4

4. The factors of $2x^2 - 8$ are:

    a. $2(4x^2)$
    b. $2(x^2 + 4)$
    c. $2(x + 2)(x + 2)$
    d. $2(x + 2)(x - 2)$

5. Two of the interior angles of a triangle are 35° and 70°. What is the measure of the last interior angle?

    a. 60°
    b. 75°
    c. 90°
    d. 100°

6. A square field has an area of 400 square feet. What is its perimeter?

    a. 100 feet
    b. 80 feet
    c. $40\sqrt{2}$ feet
    d. 40 feet

7. $\frac{5}{3} \times \frac{7}{6} =$

   a. $\frac{3}{5}$

   b. $\frac{18}{3}$

   c. $\frac{45}{31}$

   d. $\frac{35}{18}$

8. One apple costs $2. One papaya costs $3. If Samantha spends $35 and gets 15 pieces of fruit, how many papayas did she buy?

   a. Three
   b. Four
   c. Five
   d. Six

9. If $x^2 - 6 = 30$, then one possible value for x is:

   a. -6
   b. -4
   c. 3
   d. 5

10. A cube has a side length of 6 inches. What is its volume?

   a. 6 cubic inches
   b. 36 cubic inches
   c. 144 cubic inches
   d. 216 cubic inches

11. A square has a side length of 4 inches. A triangle has a base of 2 inches and a height of 8 inches. What is the total area of the square and triangle?

   a. 24 square inches
   b. 28 square inches
   c. 32 square inches
   d. 36 square inches

12. $-\frac{1}{3}\sqrt{81} =$

   a. -9
   b. -3
   c. 0
   d. +9

13. Simplify $(2x - 3)(4x + 2)$

   a. $8x^2 - 8x - 6$
   b. $6x^2 + 8x - 5$
   c. $-4x^2 - 8x - 1$
   d. $4x^2 - 4x - 6$

14. $\frac{11}{6} - \frac{3}{8} =$

    a. $\frac{5}{4}$

    b. $\frac{51}{36}$

    c. $\frac{35}{24}$

    d. $\frac{3}{2}$

15. A triangle is to have a base $\frac{1}{3}$ as long as its height. Its area must be 6 square feet. How long will its base be?

    a. 1 foot

    b. 1.5 feet

    c. 2 feet

    d. 2.5 feet

16. Which is closest to 17.8×9.9?

    a. 140

    b. 180

    c. 200

    d. 350

17. 6 is 30% of what number?

    a. 18

    b. 22

    c. 24

    d. 20

18. $3\frac{2}{3} - 1\frac{4}{5} =$

    a. $1\frac{13}{15}$

    b. $\frac{14}{15}$

    c. $2\frac{2}{3}$

    d. $\frac{4}{5}$

19. What is $\frac{420}{98}$ rounded to the nearest integer?

    a. 7

    b. 3

    c. 5

    d. 4

20. Which of the following is largest?
    a. 0.45
    b. 0.096
    c. 0.3
    d. 0.313

21. What is the value of b in this equation?
$$5b - 4 = 2b + 17$$

    a. 13
    b. 24
    c. 7
    d. 21

22. Twenty is 40% of what number?
    a. 50
    b. 8
    c. 200
    d. 5000

23. Which of the following expressions is equivalent to this equation?
$$\frac{2xy^2 + 4x - 8y}{16xy}$$

    a. $\frac{y}{8} + \frac{1}{4y} - \frac{1}{2x}$
    b. $8xy + 4y - 2x$
    c. $xy^2 + \frac{x}{4y} - \frac{1}{2x}$
    d. $\frac{y}{8} + 4y - 8y$

24. Arrange the following numbers from least to greatest value:

$0.85, \frac{4}{5}, \frac{2}{3}, \frac{91}{100}$

    a. $0.85, \frac{4}{5}, \frac{2}{3}, \frac{91}{100}$

    b. $\frac{4}{5}, 0.85, \frac{91}{100}, \frac{2}{3}$

    c. $\frac{2}{3}, \frac{4}{5}, 0.85, \frac{91}{100}$

    d. $0.85, \frac{91}{100}, \frac{4}{5}, \frac{2}{3}$

25. Simplify the following expression:

$$(3x + 5)(x - 8)$$

    a. $3x^2 - 19x - 40$
    b. $4x - 19x - 13$
    c. $3x^2 - 19x + 40$
    d. $3x^2 + 5x - 3$

26. If $6t + 4 = 16$, what is t?
    a. 1
    b. 2
    c. 3
    d. 4

27. The variable y is directly proportional to x. If $y = 3$ when $x = 5$, then what is y when $x = 20$?
    a. 10
    b. 12
    c. 14
    d. 16

28. A line passes through the point (1, 2) and crosses the y-axis at $y = 1$. Which of the following is an equation for this line?
    a. $y = 2x$
    b. $y = x + 1$
    c. $x + y = 1$
    d. $y = \frac{x}{2} - 2$

29. There are $4x + 1$ treats in each party favor bag. If a total of $60x + 15$ treats are distributed, how many bags are given out?
    a. 15
    b. 16
    c. 20
    d. 22

30. Apples cost $2 each, while bananas cost $3 each. Maria purchased 10 fruits in total and spent $22. How many apples did she buy?
    a. 5
    b. 6
    c. 7
    d. 8

Answer Explanations

1. D: Start by taking a common denominator of 30.

$$\frac{14}{15} = \frac{28}{30}$$

$$\frac{3}{5} = \frac{18}{30}$$

$$\frac{1}{30} = \frac{1}{30}$$

Add and subtract the numerators for the next step.

$$\frac{28}{30} + \frac{18}{30} - \frac{1}{30} = \frac{28 + 18 - 1}{30} = \frac{45}{30} = \frac{3}{2}$$

In the last step, the 15 is factored out from the numerator and denominator.

2. A: From the second equation, add x to both sides and subtract 1 from both sides:

$$-x + 3y + x - 1 = 1 + x - 1$$

with the result of $3y - 1 = x$.

Substitute this into the first equation and get:

$$3(3y - 1) + 2y = 8$$

$$9y - 3 + 2y = 8$$

$$11y = 11$$

$$y = 1$$

Putting this into $3y - 1 = x$ gives:

$$3(1) - 1 = x \text{ or } x = 2, y = 1$$

3. C: First, the square root of 16 is 4. So this simplifies to:

$$\frac{1}{2}\sqrt{16}$$

$$\frac{1}{2}(\pm 4) = \pm 2$$

Only +2 shows up in the answer choices, so that one is correct.

4. D: The easiest way to approach this problem is to factor out a 2 from each term.

$$2x^2 - 8 = 2(x^2 - 4)$$

Use the formula:

$$x^2 - y^2 = (x + y)(x - y) \text{ to factor:}$$

$$x^2 - 4$$

$$x^2 - 2^2$$

$$(x + 2)(x - 2)$$

So:

$$2(x^2 - 4) = 2(x + 2)(x - 2)$$

5. B: The total of the interior angles of a triangle must be 180°. The sum of the first two is 105°, so the remaining is $180° - 105° = 75°$.

6. B: The length of the side will be $\sqrt{400}$. The calculation is performed a bit more easily by breaking this into the product of two square roots:

$$\sqrt{400} = \sqrt{4 \times 100}$$

$$\sqrt{4} \times \sqrt{100} = 2 \times 10$$

$$20 \text{ feet}$$

However, there are 4 sides, so the total is:

$$20 \times 4 = 80 \text{ feet}$$

7. D: To take the product of two fractions, just multiply the numerators and denominators.

$$\frac{5}{3} \times \frac{7}{6} = \frac{5 \times 7}{3 \times 6} = \frac{35}{18}$$

The numerator and denominator have no common factors, so this is simplified completely.

8. C: Let a be the number of apples purchased, and let p be the number of papayas purchased. There is a total of 15 pieces of fruit, so one equation is:

$$a + p = 15$$

The total cost is $35, and in terms of the total apples and papayas purchased as:

$$2a + 3p = 35$$

If we multiply the first equation by 2 on both sides, it becomes:

$$2a + 2p = 30$$

We then subtract this equation from the second equation:

$$2a + 3p - (2a + 2p) = 35 - 30$$

$$p = 5$$

So, five papayas were purchased.

9. A: This equation can be solved as follows: $x^2 = 36$, so $x = \pm\sqrt{36} = \pm 6$. Only -6 shows up in the list.

10. D: The volume of a cube is given by cubing the length of its side. $6^3 = 6 \times 6 \times 6 = 36 \times 6 = 216$.

11. A: The area of the square is the square of its side length, so $4^2 = 16$ square inches. The area of a triangle is half the base times the height, so $\frac{1}{2} \times 2 \times 8 = 8$ square inches. The total is $16 + 8 = 24$ square inches.

12. B:

$$-\frac{1}{3}\sqrt{81}$$

$$-\frac{1}{3} \pm (9) = \pm 3$$

Only -3 shows up in the answer choices, so that one is correct.

13. A: Multiply each of the terms in the first parentheses and then multiply each of the terms in the second parentheses.

$$(2x - 3)(4x + 2)$$

$$2x(4x) + 2x(2) - 3(4x) - 3(2)$$

$$8x^2 + 4x - 12x - 6 = 8x^2 - 8x - 6$$

14. C: To solve, make the denominator equal 24, since that is the lowest common multiple of 6 and 8.

$$\frac{11}{6} - \frac{3}{8}$$

$$\frac{11}{6} \times \frac{4}{4} = \frac{44}{24}$$

$$\frac{3}{8} \times \frac{3}{3} = \frac{9}{24}$$

$$\frac{44}{24} - \frac{9}{24}$$

$$\frac{44 - 9}{24} = \frac{35}{24}$$

15. C: The formula for the area of a triangle with base b and height h is $\frac{1}{2}bh$, where the base is one-third the height, or $b = \frac{1}{3}h$ or equivalently $h = 3b$. Using the formula for a triangle, this becomes:

$$\frac{1}{2}b(3b) = \frac{3}{2}b^2$$

Now, this has to be equal to 6. So:

$$\frac{3}{2}b^2 = 6$$

$$b^2 = 4$$

$$b = \pm 2$$

However, lengths are positive, so the base must be 2 feet long.

16. B: Instead of multiplying these out, the product can be estimated by using $18 \times 10 = 180$. The error here should be lower than 15, since it is rounded to the nearest integer, and the numbers add to something less than 30.

17. D: 30% is $\frac{3}{10}$. The number itself must be $\frac{10}{3}$ of 6, or:

$$\frac{10}{3} \times 6$$

$$10 \times 2 = 20$$

18. A: First, these numbers need to be converted to improper fractions: $\frac{11}{3} - \frac{9}{5}$. Take 15 as a common denominator:

$$\frac{11}{3} - \frac{9}{5}$$

$$\frac{55}{15} - \frac{27}{15}$$

$$\frac{28}{15} = 1\frac{13}{15}$$

19. D: Dividing by 98 can be approximated by dividing by 100, which would mean shifting the decimal point of the numerator to the left by 2. The result is 4.2 and rounds to 4.

20. A: To figure out which is largest, look at the first non-zero digits. Answer *B's* first nonzero digit is in the hundredths place. The other three all have nonzero digits in the tenths place, so it must be *A, C,* or *D*. Of these, *A* has the largest first nonzero digit.

21. C: To solve for the value of b, both sides of the equation need to be equalized.

Start by cancelling out the lower value of -4 by adding 4 to both sides:

$$5b - 4 = 2b + 17$$
$$5b - 4 + 4 = 2b + 17 + 4$$
$$5b = 2b + 21$$

The variable *b* is the same on each side, so subtract the lower 2b from each side:

$$5b = 2b + 21$$
$$5b - 2b = 2b + 21 - 2b$$
$$3b = 21$$

Then divide both sides by 3 to get the value of *b*:

$$3b = 21$$

$$\frac{3b}{3} = \frac{21}{3}$$

$$b = 7$$

22. A: Setting up a proportion is the easiest way to represent this situation. The proportion becomes $\frac{20}{x} = \frac{40}{100}$, where cross-multiplication can be used to solve for x. Here, $40x = 2000$, so $x = 50$.

23. A: First, separate each element of the numerator with the denominator as follows:

$$\frac{2xy^2}{16xy} + \frac{4x}{16xy} - \frac{8y}{16xy}$$

Simplify each expression accordingly, reaching answer *A*:

$$\frac{y}{8} + \frac{1}{4y} - \frac{1}{2x}$$

24. C: The first step is to depict each number using decimals.

$$\frac{91}{100} = 0.91$$

Dividing the numerator by denominator of $\frac{4}{5}$ to convert it to a decimal yields 0.80, while $\frac{2}{3}$ becomes 0.66 recurring. Rearrange each expression in ascending order, as found in answer *C*.

25. A: When parentheses are around two expressions, they need to be *multiplied*. In this case, separate each expression into its parts (separated by addition and subtraction) and multiply by each of the parts in the other expression. Then, add the products together.

$$(3x)(x) + (3x)(-8) + (+5)(x) + (+5)(-8)$$

$$3x^2 - 24x + 5x - 40$$

Remember that when multiplying a positive integer by a negative integer, it will remain negative. Then add $-24x + 5x$ to get the simplified expression, answer A.

26. B: First, subtract 4 from each side. This yields $6t = 12$. Now, divide both sides by 6 to obtain $t = 2$.

27. B: To be directly proportional means that $y = mx$. If x is changed from 5 to 20, the value of x is multiplied by 4. Applying the same rule to the y-value, also multiply the value of y by 4. Therefore, $y = 12$.

28. B: From the slope-intercept form, $y = mx + b$, it is known that b is the y-intercept, which is 1. Compute the slope as $\frac{2-1}{1-0} = 1$, so the equation should be $y = x + 1$.

29. A: Each bag contributes $4x + 1$ treats. The total treats will be in the form $4nx + n$ where n is the total number of bags. The total is in the form $60x + 15$, from which it is known $n = 15$.

30. D: Let a be the number of apples and b the number of bananas. Then, the total cost is:

$$2a + 3b = 22$$

While it also known that:

$$a + b = 10$$

Using the knowledge of systems of equations, cancel the b variables by multiplying the second equation by -3. This makes the equation:

$$-3a - 3b = -30$$

Adding this to the first equation, the o values cancel to get $-a = -8$, which simplifies to $a = 8$.

Reading Comprehension

Literary Analysis

Style, Tone, and Mood

Style, tone, and mood are often thought to be the same thing. Though they're closely related, there are important differences to keep in mind. The easiest way to do this is to remember that style "creates and affects" tone and mood. More specifically, style is how the writer uses words to create the desired tone and mood for their writing.

Style
Style can include any number of technical writing choices. A few examples of style choices include:

- Sentence Construction: When presenting facts, does the writer use shorter sentences to create a quicker sense of the supporting evidence, or do they use longer sentences to elaborate and explain the information?

- Technical Language: Does the writer use jargon to demonstrate their expertise in the subject, or do they use ordinary language to help the reader understand things in simple terms?

- Formal Language: Does the writer refrain from using contractions such as *won't* or *can't* to create a more formal tone, or do they use a colloquial, conversational style to connect to the reader?

- Formatting: Does the writer use a series of shorter paragraphs to help the reader follow a line of argument, or do they use longer paragraphs to examine an issue in great detail and demonstrate their knowledge of the topic?

On the test, examine the writer's style and how their writing choices affect the way the text comes across.

Tone
Tone refers to the writer's attitude toward the subject matter. Tone is usually explained in terms of a work of fiction. For example, the tone conveys how the writer feels about their characters and the situations in which they're involved. Nonfiction writing is sometimes thought to have no tone at all; however, this is incorrect.

A lot of nonfiction writing has a neutral tone, which is an important tone for the writer to take. A neutral tone demonstrates that the writer is presenting a topic impartially and letting the information speak for itself. On the other hand, nonfiction writing can be just as effective and appropriate if the tone isn't neutral. For instance, take this example involving seat belts:

> Seat belts save more lives than any other automobile safety feature. Many studies show that airbags save lives as well; however, not all cars have airbags. For instance, some older cars don't. Furthermore, air bags aren't entirely reliable. For example, studies show that in 15% of accidents airbags don't deploy as designed, but, on the other hand, seat belt malfunctions are extremely rare. The number of highway fatalities has plummeted since laws requiring seat belt usage were enacted.

In this passage, the writer mostly chooses to retain a neutral tone when presenting information. If the writer would instead include their own personal experience of losing a friend or family member in a car accident, the tone would change dramatically. The tone would no longer be neutral and would show that the writer has a personal stake in the content, allowing them to interpret the information in a different way. When analyzing tone, consider what the writer is trying to achieve in the text and how they *create* the tone using style.

Mood
Mood refers to the feelings and atmosphere that the writer's words create for the reader. Like tone, many nonfiction texts can have a neutral mood. To return to the previous example, if the writer would choose to include information about a person they know being killed in a car accident, the text would suddenly carry an emotional component that is absent in the previous example. Depending on how they present the information, the writer can create a sad, angry, or even hopeful mood. When analyzing the mood, consider what the writer wants to accomplish and whether the best choice was made to achieve that end.

Consistency

Whatever style, tone, and mood the writer uses, good writing should remain consistent throughout. If the writer chooses to include the tragic, personal experience above, it would affect the style, tone, and mood of the entire text. It would seem out of place for such an example to be used in the middle of a neutral, measured, and analytical text. To adjust the rest of the text, the writer needs to make additional choices to remain consistent. For example, the writer might decide to use the word *tragedy* in place of the more neutral *fatality*, or they could describe a series of car-related deaths as an *epidemic*. Adverbs and adjectives such as *devastating* or *horribly* could be included to maintain this consistent attitude toward the content. When analyzing writing, look for sudden shifts in style, tone, and mood, and consider whether the writer would be wiser to maintain the prevailing strategy.

Identify the Position and Purpose

When it comes to an author's writing, readers should always identify a position or stance. No matter how objective a text may seem, readers should assume the author has preconceived beliefs. One can reduce the likelihood of accepting an invalid argument by looking for multiple articles on the topic, including those with varying opinions. If several opinions point in the same direction and are backed by reputable peer-reviewed sources, it's more likely the author has a valid argument. Positions that run contrary to widely held beliefs and existing data should invite scrutiny. There are exceptions to the rule, so be a careful consumer of information.

Though themes, symbols, and motifs are buried deep within the text and can sometimes be difficult to infer, an author's purpose is usually obvious from the beginning. There are four purposes of writing: to inform, to persuade, to describe, and to entertain. Informative writing presents facts in an accessible way. Persuasive writing appeals to emotions and logic to inspire the reader to adopt a specific stance. Be wary of this type of writing, as it can mask a lack of objectivity with powerful emotion. Descriptive writing is designed to paint a picture in the reader's mind, while texts that entertain are often narratives designed to engage and delight the reader.

The various writing styles are usually blended, with one purpose dominating the rest. A persuasive text, for example, might begin with a humorous tale to make readers more receptive to the persuasive message, or a recipe in a cookbook designed to inform might be preceded by an entertaining anecdote that makes the recipes more appealing.

Identify Passage Characteristics

Writing can be classified under four passage types: narrative, expository, descriptive (sometimes called technical), and persuasive. Though these types are not mutually exclusive, one form tends to dominate the rest. By recognizing the *type* of passage you're reading, you gain insight into *how* you should read. When reading a narrative intended to entertain, sometimes you can read more quickly through the passage if the details are discernible. A technical document, on the other hand, might require a close read, because skimming the passage might cause the reader to miss salient details.

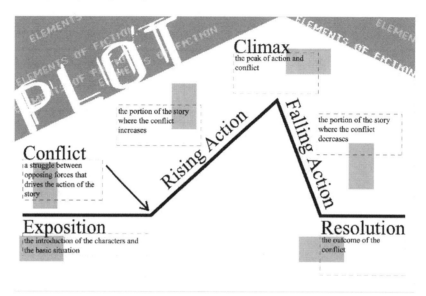

1. Narrative writing, at its core, is the art of storytelling. For a narrative to exist, certain elements must be present. It must have characters. While many characters are human, characters could be defined as anything that thinks, acts, and talks like a human. For example, many recent movies, such as *Lord of the Rings* and *The Chronicles of Narnia*, include animals, fantasy creatures, and even trees that behave like humans. Narratives also must have a plot or sequence of events. Typically, those events follow a standard plot diagram, but recent trends start *in medias res* or in the middle (nearer the climax). In this instance, foreshadowing and flashbacks often fill in plot details. Along with characters and a plot, there must also be conflict. Conflict is usually divided into two types: internal and external. Internal conflict indicates the character is in turmoil. Think of an angel on one shoulder and the devil on the other, arguing it out. Internal conflicts are presented through the character's thoughts. External conflicts are visible. Types of external conflict include person versus person, person versus nature, person versus technology, person versus the supernatural, or a person versus fate.

2. Expository writing is detached and to the point, while other types of writing — persuasive, narrative, and descriptive — are livelier. Since expository writing is designed to instruct or inform, it usually involves directions and steps written in second person ("you" voice) and lacks any persuasive or narrative elements. Sequence words such as *first*, *second*, and *third*, or *in the first place*, *secondly*, and *lastly* are often given to add fluency and cohesion. Common examples of expository writing include instructor's lessons, cookbook recipes, and repair manuals.

3. Due to its empirical nature, technical writing is filled with steps, charts, graphs, data, and statistics. The goal of technical writing is to advance understanding in a field through the scientific method. Experts such as teachers, doctors, or mechanics use words unique to the profession in which they

operate. These words, which often incorporate acronyms, are called *jargon*. Technical writing is a type of expository writing but is not meant to be understood by the general public. Instead, technical writers assume readers have received a formal education in a particular field of study and need no explanation as to what the jargon means. Imagine a doctor trying to understand a diagnostic reading for a car or a mechanic trying to interpret lab results. Only professionals with proper training will fully comprehend the text.

4. Persuasive writing is designed to change opinions and attitudes. The topic, stance, and arguments are found in the thesis, positioned near the end of the introduction. Later supporting paragraphs offer relevant quotations, paraphrases, and summaries from primary or secondary sources, which are then interpreted, analyzed, and evaluated. The goal of persuasive writers is not to stack quotes, but to develop original ideas by using sources as a starting point. Good persuasive writing makes powerful arguments with valid sources and thoughtful analysis. Poor persuasive writing is riddled with bias and logical fallacies. Sometimes, logical and illogical arguments are sandwiched together in the same text. Therefore, readers should display skepticism when reading persuasive arguments.

Interpret Influences of Historical Context

Studying historical literature is fascinating. It reveals a snapshot in time of people, places, and cultures; a collective set of beliefs and attitudes that no longer exist. Writing changes as attitudes and cultures evolve. Beliefs previously considered immoral or wrong may be considered acceptable today. Researching the historical period of an author gives the reader perspective. The dialogue in Jane Austen's *Pride and Prejudice*, for example, is indicative of social class during the Regency era. Similarly, the stereotypes and slurs in *The Adventures of Huckleberry Finn* were a result of common attitudes and beliefs in the late 1800s, attitudes now found to be reprehensible.

Recognizing Cultural Themes

Regardless of culture, place, or time, certain themes are universal to the human condition. Because humans experience joy, rage, jealousy, and pride, certain themes span centuries. For example, Shakespeare's *Macbeth,* as well as modern works like *The 50th Law* by rapper 50 Cent and Robert Greene or the Netflix series *House of Cards* all feature characters who commit atrocious acts because of ambition. Similarly, *The Adventures of Huckleberry Finn*, published in the 1880s, and *The Catcher in the Rye*, published in the 1950s, both have characters who lie, connive, and survive on their wits.

Moviegoers know whether they are seeing an action, romance or horror film, and are often disappointed if the movie doesn't fit into the conventions of a particular category. Similarly, categories or genres give readers a sense of what to expect from a text. Some of the most basic genres in literature include books, short stories, poetry, and drama. Many genres can be split into sub-genres. For example, the sub-genres of historical fiction, realistic fiction, and fantasy all fit under the fiction genre.

Each genre has a unique way of approaching a particular theme. Books and short stories use plot, characterization, and setting, while poems rely on figurative language, sound devices, and symbolism. Dramas reveal plot through dialogue and the actor's voice and body language.

Paragraph Comprehension

Topic Versus the Main Idea

It is very important to know the difference between the topic and the main idea of the text. Even though these two are similar because they both present the central point of a text, they have distinctive differences. A *topic* is the subject of the text; it can usually be described in a one- to two-word phrase and appears in the simplest form. On the other hand, the *main idea* is more detailed and provides the author's central point of the text. It can be expressed through a complete sentence and can be found in the beginning, middle, or end of a paragraph. In most nonfiction books, the first sentence of the passage usually (but not always) states the main idea. Take a look at the passage below to review the topic versus the main idea.

Cheetahs

Cheetahs are one of the fastest mammals on land, reaching up to 70 miles an hour over short distances. Even though cheetahs can run as fast as 70 miles an hour, they usually only have to run half that speed to catch up with their choice of prey. Cheetahs cannot maintain a fast pace over long periods of time because they will overheat their bodies. After a chase, cheetahs need to rest for approximately 30 minutes prior to eating or returning to any other activity.

In the example above, the topic of the passage is "Cheetahs" simply because that is the subject of the text. The main idea of the text is "Cheetahs are one of the fastest mammals on land but can only maintain this fast pace for short distances." While it covers the topic, it is more detailed and refers to the text in its entirety. The text continues to provide additional details called *supporting details,* which will be discussed in the next section.

Supporting Details

Supporting details help readers better develop and understand the main idea. Supporting details answer questions like *who, what, where, when, why,* and *how.* Different types of supporting details include examples, facts and statistics, anecdotes, and sensory details.

Persuasive and informative texts often use supporting details. In persuasive texts, authors attempt to make readers agree with their point of view, and supporting details are often used as "selling points." If authors make a statement, they should support the statement with evidence in order to adequately persuade readers. Informative texts use supporting details such as examples and facts to inform readers. Take another look at the previous "Cheetahs" passage to find examples of supporting details.

Cheetahs

Cheetahs are one of the fastest mammals on land, reaching up to 70 miles an hour over short distances. Even though cheetahs can run as fast as 70 miles an hour, they usually only have to run half that speed to catch up with their choice of prey. Cheetahs cannot maintain a fast pace over long periods of time because they will overheat their bodies. After a chase, cheetahs need to rest for approximately 30 minutes prior to eating or returning to any other activity.

In the example above, supporting details include:

- Cheetahs reach up to 70 miles per hour over short distances.
- They usually only have to run half that speed to catch up with their prey.
- Cheetahs will overheat their bodies if they exert a high speed over longer distances.
- Cheetahs need to rest for 30 minutes after a chase.

Look at the diagram below (applying the cheetah example) to help determine the hierarchy of topic, main idea, and supporting details.

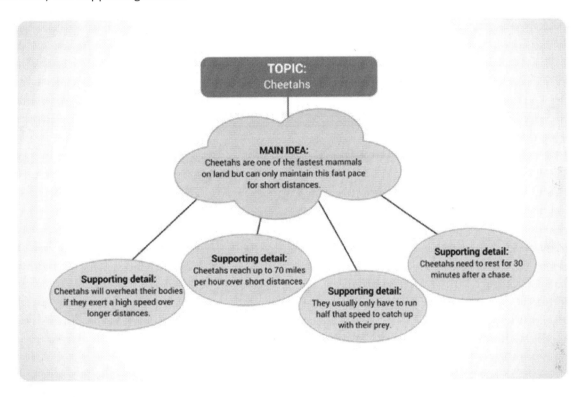

Drawing Conclusions

When drawing conclusions about texts or passages, readers should do two main things: 1) Use the information that they already know and 2) Use the information they have learned from the text or passage. Authors write with an intended purpose, and it is the reader's responsibility to understand and form logical conclusions of authors' ideas. It is important to remember that the reader's conclusions should be supported by information directly from the text. Readers cannot simply form conclusions based off of only information they already know.

There are several ways readers can draw conclusions from authors' ideas, such as note taking, text evidence, text credibility, writing a response to text, directly stated information versus implications, outlining, summarizing, and paraphrasing. Let's take a look at each important strategy to help readers draw logical conclusions.

Note Taking

When readers take notes throughout texts or passages, they are jotting down important facts or points that the author makes. Note taking is a useful record of information that helps readers understand the text or passage and respond to it. When taking notes, readers should keep lines brief and filled with pertinent information so that they are not rereading a large amount of text, but rather just key points, elements, or words. After readers have completed a text or passage, they can refer to their notes to help them form a conclusion about the author's ideas in the text or passage.

Text Evidence

Text evidence is the information readers find in a text or passage that supports the main idea or point(s) in a story. In turn, text evidence can help readers draw conclusions about the text or passage. The information should be taken directly from the text or passage and placed in quotation marks. Text evidence provides readers with information to support ideas about the text so that they do not rely simply on their own thoughts. Details should be precise, descriptive, and factual. Statistics are a great piece of text evidence because they provide readers with exact numbers and not just a generalization. For example, instead of saying "Asia has a larger population than Europe," authors could provide detailed information such as, "In Asia there are over 4 billion people, whereas in Europe there are a little over 750 million." More definitive information provides better evidence to readers to help support their conclusions about texts or passages.

Text Credibility

Credible sources are important when drawing conclusions because readers need to be able to trust what they are reading. Authors should always use credible sources to help gain the trust of their readers. A text is *credible* when it is believable and the author is objective and unbiased. If readers do not trust an author's words, they may simply dismiss the text completely. For example, if an author writes a persuasive essay, he or she is outwardly trying to sway readers' opinions to align with his or her own. Readers may agree or disagree with the author, which may, in turn, lead them to believe that the author is credible or not credible. Also, readers should keep in mind the source of the text. If readers review a journal about astronomy, would a more reliable source be a NASA employee or a medical doctor? Overall, text credibility is important when drawing conclusions, because readers want reliable sources that support the decisions they have made about the author's ideas.

Writing a Response to Text

Once readers have determined their opinions and validated the credibility of a text, they can then reflect on the text. Writing a response to a text is one way readers can reflect on the given text or passage. When readers write responses to a text, it is important for them to rely on the evidence within the text to support their opinions or thoughts. Supporting evidence such as facts, details, statistics, and quotes directly from the text are key pieces of information readers should reflect upon or use when writing a response to text.

Directly Stated Information Versus Implications

Engaged readers should constantly self-question while reviewing texts to help them form conclusions. Self-questioning is when readers review a paragraph, page, passage, or chapter and ask themselves, "Did I understand what I read?," "What was the main event in this section?," "Where is this taking place?," and so on. Authors can provide clues or pieces of evidence throughout a text or passage to guide readers toward a conclusion. This is why active and engaged readers should read the text or passage in its entirety before forming a definitive conclusion. If readers do not gather all the pieces of evidence needed, then they may jump to an illogical conclusion.

At times, authors directly state conclusions while others simply imply them. Of course, it is easier if authors outwardly provide conclusions to readers, because it does not leave any information open to interpretation. On the other hand, implications are things that authors do not directly state but can be assumed based off of information they provided. If authors only imply what may have happened, readers can form a menagerie of ideas for conclusions. For example, look at the following statement: "Once we heard the sirens, we hunkered down in the storm shelter." In this statement, the author does not directly state that there was a tornado, but clues such as "sirens" and "storm shelter" provide insight to the readers to help form that conclusion.

Outlining

An outline is a system used to organize writing. When reading texts, outlining is important because it helps readers organize important information in a logical pattern using roman numerals. Usually, outlines start with the main idea(s) and then branch out into subgroups or subsidiary thoughts of subjects. Not only do outlines provide a visual tool for readers to reflect on how events, characters, settings, or other key parts of the text or passage relate to one another, but they can also lead readers to a stronger conclusion.

The sample below demonstrates what a general outline looks like.

I.   Main Topic 1
     a. Subtopic 1
     b. Subtopic 2
         1. Detail 1
         2. Detail 1
II. Main Topic 2
     a. Subtopic 1
     b. Subtopic 2
         1. Detail 1
         2. Detail 2

Summarizing

At the end of a text or passage, it is important to summarize what the readers read. Summarizing is a strategy in which readers determine what is important throughout the text or passage, shorten those ideas, and rewrite or retell it in their own words. A summary should identify the main idea of the text or passage. Important details or supportive evidence should also be accurately reported in the summary. If writers provide irrelevant details in the summary, it may cloud the greater meaning of the summary in the text. When summarizing, writers should not include their opinions, quotes, or what they thought the author should have said. A clear summary provides clarity of the text or passage to the readers. Let's review the checklist of items writers should include in their summary.

Summary Checklist
- Title of the story
- Someone: Who is or are the main character(s)?
- Wanted: What did the character(s) want?
- But: What was the problem?
- So: How did the character(s) solve the problem?
- Then: How did the story end? What was the resolution?

Paraphrasing

Another strategy readers can use to help them fully comprehend a text or passage is paraphrasing. Paraphrasing is when readers take the author's words and put them into their own words. When readers and writers paraphrase, they should avoid copying the text—that is plagiarism. It is also important to include as many details as possible when restating the facts. Not only will this help readers and writers recall information, but by putting the information into their own words, they demonstrate whether or not they fully comprehend the text or passage. Look at the example below showing an original text and how to paraphrase it.

> *Original Text*: Fenway Park is home to the beloved Boston Red Sox. The stadium opened on April 20, 1912. The stadium currently seats over 37,000 fans, many of whom travel from all over the country to experience the iconic team and nostalgia of Fenway Park.

> *Paraphrased*: On April 20, 1912, Fenway Park opened. Home to the Boston Red Sox, the stadium now seats over 37,000 fans. Many spectators travel to watch the Red Sox and experience the spirit of Fenway Park.

Paraphrasing, summarizing, and quoting can often cross paths with one another. Review the chart below showing the similarities and differences between the three strategies.

Paraphrasing	Summarizing	Quoting
Uses own words	Puts main ideas into own words	Uses words that are identical to text
References original source	References original source	Requires quotation marks
Uses own sentences	Shows important ideas of source	Uses author's own words and ideas

Inferences in a Text

Readers should be able to make *inferences*. Making an inference requires the reader to read between the lines and look for what is *implied* rather than what is directly stated. That is, using information that is known from the text, the reader is able to make a logical assumption about information that is *not* directly stated but is probably true. Read the following passage:

> "Hey, do you wanna meet my new puppy?" Jonathan asked.

> "Oh, I'm sorry but please don't—" Jacinta began to protest, but before she could finish, Jonathan had already opened the passenger side door of his car and a perfect white ball of fur came bouncing towards Jacinta.

> "Isn't he the cutest?" beamed Jonathan.

> "Yes—achoo!—he's pretty—aaaachooo!!—adora—aaa—aaaachoo!" Jacinta managed to say in between sneezes. "But if you don't mind, I—I—achoo!—need to go inside."

Which of the following can be inferred from Jacinta's reaction to the puppy?
    a. she hates animals
    b. she is allergic to dogs
    c. she prefers cats to dogs
    d. she is angry at Jonathan

An inference requires the reader to consider the information presented and then form their own idea about what is probably true. Based on the details in the passage, what is the best answer to the question? Important details to pay attention to include the tone of Jacinta's dialogue, which is overall polite and apologetic, as well as her reaction itself, which is a long string of sneezes. Answer choices (a) and (d) both express strong emotions ("hates" and "angry") that are not evident in Jacinta's speech or actions. Answer choice (c) mentions cats, but there is nothing in the passage to indicate Jacinta's feelings about cats. Answer choice (b), "she is allergic to dogs," is the most logical choice—based on the fact that she began sneezing as soon as a fluffy dog approached her, it makes sense to guess that Jacinta might be allergic to dogs. So even though Jacinta never directly states, "Sorry, I'm allergic to dogs!" using the clues in the passage, it is still reasonable to guess that this is true.

Making inferences is crucial for readers of literature, because literary texts often avoid presenting complete and direct information to readers about characters' thoughts or feelings, or they present this information in an unclear way, leaving it up to the reader to interpret clues given in the text. In order to make inferences while reading, readers should ask themselves:

- What details are being presented in the text?
- Is there any important information that seems to be missing?
- Based on the information that the author *does* include, what else is probably true?
- Is this inference reasonable based on what is already known?

Apply Information

A natural extension of being able to make an inference from a given set of information is also being able to apply that information to a new context. This is especially useful in non-fiction or informative writing. Considering the facts and details presented in the text, readers should consider how the same information might be relevant in a different situation. The following is an example of applying an inferential conclusion to a different context:

> Often, individuals behave differently in large groups than they do as individuals. One example of this is the psychological phenomenon known as the bystander effect. According to the bystander effect, the more people who witness an accident or crime occur, the less likely each individual bystander is to respond or offer assistance to the victim. A classic example of this is the murder of Kitty Genovese in New York City in the 1960s. Although there were over thirty witnesses to her killing by a stabber, none of them intervened to help Kitty or contact the police.

Considering the phenomenon of the bystander effect, what would probably happen if somebody tripped on the stairs in a crowded subway station?
    a. Everybody would stop to help the person who tripped
    b. Bystanders would point and laugh at the person who tripped
    c. Someone would call the police after walking away from the station
    d. Few if any bystanders would offer assistance to the person who tripped

This question asks readers to apply the information they learned from the passage, which is an informative paragraph about the bystander effect. According to the passage, this is a concept in psychology that describes the way people in groups respond to an accident—the more people are present, the less likely any one person is to intervene. While the passage illustrates this effect with the example of a woman's murder, the question asks readers to apply it to a different context—in this case, someone falling down the stairs in front of many subway passengers. Although this specific situation is not discussed in the passage, readers should be able to apply the general concepts described in the paragraph. The definition of the bystander effect includes any instance of an accident or crime in front of a large group of people. The question asks about a situation that falls within the same definition, so the general concept should still hold true: in the midst of a large crowd, few individuals are likely to actually respond to an accident. In this case, answer choice (d) is the best response.

Author's Use of Language

Authors utilize a wide range of techniques to tell a story or communicate information. Readers should be familiar with the most common of these techniques. Techniques of writing are also commonly known as rhetorical devices.

Types of Appeals

In non-fiction writing, authors employ argumentative techniques to present their opinion to readers in the most convincing way. First of all, persuasive writing usually includes at least one type of appeal: an appeal to logic (logos), emotion (pathos), or credibility and trustworthiness (ethos). When a writer appeals to logic, they are asking readers to agree with them based on research, evidence, and an established line of reasoning. An author's argument might also appeal to readers' emotions, perhaps by including personal stories and anecdotes (a short narrative of a specific event). A final type of appeal, appeal to authority, asks the reader to agree with the author's argument on the basis of their expertise or credentials. Consider three different approaches to arguing the same opinion:

Logic (Logos)
This is an example of an appeal to logic:

> Our school should abolish its current ban on cell phone use on campus. This rule was adopted last year as an attempt to reduce class disruptions and help students focus more on their lessons. However, since the rule was enacted, there has been no change in the number of disciplinary problems in class. Therefore, the rule is ineffective and should be done away with.

The author uses evidence to disprove the logic of the school's rule (the rule was supposed to reduce discipline problems; the number of problems has not been reduced; therefore, the rule is not working) and call for its repeal.

Emotion (Pathos)
An author's argument might also appeal to readers' emotions, perhaps by including personal stories and anecdotes. The next example presents an appeal to emotion. By sharing the personal anecdote of one student and speaking about emotional topics like family relationships, the author invokes the reader's empathy in asking them to reconsider the school rule.

> Our school should abolish its current ban on cell phone use on campus. If they aren't able to use their phones during the school day, many students feel isolated from their loved ones. For

example, last semester, one student's grandmother had a heart attack in the morning. However, because he couldn't use his cell phone, the student didn't know about his grandmother's accident until the end of the day—when she had already passed away and it was too late to say goodbye. By preventing students from contacting their friends and family, our school is placing undue stress and anxiety on students.

Credibility (Ethos)

Finally, an appeal to authority includes a statement from a relevant expert. In this case, the author uses a doctor in the field of education to support the argument. All three examples begin from the same opinion—the school's phone ban needs to change—but rely on different argumentative styles to persuade the reader.

> Our school should abolish its current ban on cell phone use on campus. According to Dr. Bartholomew Everett, a leading educational expert, "Research studies show that cell phone usage has no real impact on student attentiveness. Rather, phones provide a valuable technological resource for learning. Schools need to learn how to integrate this new technology into their curriculum." Rather than banning phones altogether, our school should follow the advice of experts and allow students to use phones as part of their learning.

Rhetorical Questions

Another commonly used argumentative technique is asking rhetorical questions, questions that do not actually require an answer but that push the reader to consider the topic further.

> I wholly disagree with the proposal to ban restaurants from serving foods with high sugar and sodium contents. Do we really want to live in a world where the government can control what we eat? I prefer to make my own food choices.

Here, the author's rhetorical question prompts readers to put themselves in a hypothetical situation and imagine how they would feel about it.

Figurative Language

Literary texts also employ rhetorical devices. Figurative language like simile and metaphor is a type of rhetorical device commonly found in literature. In addition to rhetorical devices that play on the *meanings* of words, there are also rhetorical devices that use the *sounds* of words. These devices are most often found in poetry but may also be found in other types of literature and in non-fiction writing like speech texts.

Alliteration and *assonance* are both varieties of sound repetition. Other types of sound repetition include: anaphora, repetition that occurs at the beginning of the sentences; epiphora, repetition occurring at the end of phrases; antimetabole, repetition of words in reverse order; and antiphrasis, a form of denial of an assertion in a text.

Alliteration refers to the repetition of the first sound of each word. Recall Robert Burns' opening line:

> My love is like a red, red rose

This line includes two instances of alliteration: "love" and "like" (repeated *L* sound), as well as "red" and "rose" (repeated *R* sound). Next, assonance refers to the repetition of vowel sounds, and can occur anywhere within a word (not just the opening sound).

Here is the opening of a poem by John Keats:

> When I have fears that I may cease to be
>
> Before my pen has glean'd my teeming brain

Assonance can be found in the words "fears," "cease," "be," "glean'd," and "teeming," all of which stress the long *E* sound. Both alliteration and assonance create a harmony that unifies the writer's language.

Another sound device is *onomatopoeia*, or words whose spelling mimics the sound they describe. Words like "crash," "bang," and "sizzle" are all examples of onomatopoeia. Use of onomatopoetic language adds auditory imagery to the text.

Readers are probably most familiar with the technique of *pun*. A pun is a play on words, taking advantage of two words that have the same or similar pronunciation. Puns can be found throughout Shakespeare's plays, for instance:

> Now is the winter of our discontent
> Made glorious summer by this son of York

These lines from *Richard III* contain a play on words. Richard III refers to his brother, the newly crowned King Edward IV, as the "son of York," referencing their family heritage from the house of York. However, while drawing a comparison between the political climate and the weather (times of political trouble were the "winter," but now the new king brings "glorious summer"), Richard's use of the word "son" also implies another word with the same pronunciation, "sun"—so Edward IV is also like the sun, bringing light, warmth, and hope to England. Puns are a clever way for writers to suggest two meanings at once.

Counterarguments

If an author presents a differing opinion or a counterargument in order to refute it, the reader should consider how and why this information is being presented. It is meant to strengthen the original argument and shouldn't be confused with the author's intended conclusion, but it should also be considered in the reader's final evaluation.

Authors can also use bias if they ignore the opposing viewpoint or present their side in an unbalanced way. A strong argument considers the opposition and finds a way to refute it. Critical readers should look for an unfair or one-sided presentation of the argument and be skeptical, as a bias may be present. Even if this bias is unintentional, if it exists in the writing, the reader should be wary of the validity of the argument. Readers should also look for the use of stereotypes, which refer to specific groups. Stereotypes are often negative connotations about a person or place, and should always be avoided. When a critical reader finds stereotypes in a piece of writing, they should be critical of the argument, and consider the validity of anything the author presents. Stereotypes reveal a flaw in the writer's thinking and may suggest a lack of knowledge or understanding about the subject.

Meaning of Words in Context

There will be many occasions in one's reading career in which an unknown word or a word with multiple meanings will pop up. There are ways of determining what these words or phrases mean that do not require the use of the dictionary, which is especially helpful during a test where one may not be available. Even outside of the exam, knowing how to derive an understanding of a word via context clues will be a critical skill in the real world. The context is the circumstances in which a story or a passage is happening, and can usually be found in the series of words directly before or directly after the word or phrase in question. The clues are the words that hint towards the meaning of the unknown word or phrase.

There may be questions that ask about the meaning of a particular word or phrase within a passage. There are a couple ways to approach these kinds of questions:

1. Define the word or phrase in a way that is easy to comprehend (using context clues).
2. Try out each answer choice in place of the word.

To demonstrate, here's an example from *Alice in Wonderland*:

Alice was beginning to get very tired of sitting by her sister on the bank, and of having nothing to do: once or twice she <u>peeped</u> into the book her sister was reading, but it had no pictures or conversations in it, "and what is the use of a book," thought Alice, "without pictures or conversations?"

Q: As it is used in the selection, the word <u>peeped</u> means:

Using the first technique, before looking at the answers, define the word "peeped" using context clues and then find the matching answer. Then, analyze the entire passage in order to determine the meaning, not just the surrounding words.

To begin, imagine a blank where the word should be and put a synonym or definition there: "once or twice she _____ into the book her sister was reading." The context clue here is the book. It may be tempting to put "read" where the blank is, but notice the preposition word, "into." One does not read *into* a book, one simply reads a book, and since reading a book requires that it is seen with a pair of eyes, then "look" would make the most sense to put into the blank: "once or twice she <u>looked </u>into the book her sister was reading."

Once an easy-to-understand word or synonym has been supplanted, readers should check to make sure it makes sense with the rest of the passage. What happened after she looked into the book? She thought to herself how a book without pictures or conversations is useless. This situation in its entirety makes sense.

Now check the answer choices for a match:
a. To make a high-pitched cry
b. To smack
c. To look curiously
d. To pout

Since the word was already defined, Choice *C* is the best option.

Using the second technique, replace the figurative blank with each of the answer choices and determine which one is the most appropriate. Remember to look further into the passage to clarify that they work, because they could still make sense out of context.

    a. Once or twice she <u>made a high pitched cry</u> into the book her sister was reading

    b. Once or twice she <u>smacked</u> into the book her sister was reading

    c. Once or twice she <u>looked curiously</u> into the book her sister was reading

    d. Once or twice she <u>pouted</u> into the book her sister was reading

For Choice *A*, it does not make much sense in any context for a person to yell into a book, unless maybe something terrible has happened in the story. Given that afterward Alice thinks to herself how useless a book without pictures is, this option does not make sense within context.

For Choice *B*, smacking a book someone is reading may make sense if the rest of the passage indicates a reason for doing so. If Alice was angry or her sister had shoved it in her face, then maybe smacking the book would make sense within context. However, since whatever she does with the book causes her to think, "what is the use of a book without pictures or conversations?" then answer Choice *B* is not an appropriate answer. Answer Choice *C* fits well within context, given her subsequent thoughts on the matter. Answer Choice *D* does not make sense in context or grammatically, as people do not "pout into" things.

This is a simple example to illustrate the techniques outlined above. There may, however, be a question in which all of the definitions are correct and also make sense out of context, in which the appropriate context clues will really need to be honed in on in order to determine the correct answer. For example, here is another passage from *Alice in Wonderland*:

> . . . but when the Rabbit actually took a watch out of its waistcoat pocket, and looked at it, and then hurried on, Alice <u>started</u> to her feet, for it flashed across her mind that she had never before seen a rabbit with either a waistcoat-pocket or a watch to take out of it, and burning with curiosity, she ran across the field after it, and was just in time to see it pop down a large rabbit-hole under the hedge.

Q: As it is used in the passage, the word started means

    a. To turn on

    b. To begin

    c. To move quickly

    d. To be surprised

All of these words qualify as a definition of "start," but using context clues, the correct answer can be identified using one of the two techniques above. It's easy to see that one does not turn on, begin, or be surprised to one's feet. The selection also states that she "ran across the field after it," indicating that she was in a hurry. Therefore, to move quickly would make the most sense in this context.

The same strategies can be applied to vocabulary that may be completely unfamiliar. In this case, focus on the words before or after the unknown word in order to determine its definition. Take this sentence, for example:

> Sam was such a <u>miser</u> that he forced Andrew to pay him twelve cents for the candy, even though he had a large inheritance and he knew his friend was poor.

Unlike with assertion questions, for vocabulary questions, it may be necessary to apply some critical thinking skills that may not be explicitly stated within the passage. Think about the implications of the passage, or what the text is trying to say. With this example, it is important to realize that it is considered unusually stingy for a person to demand so little money from someone instead of just letting their friend have the candy, especially if this person is already wealthy. Hence, a <u>miser</u> is a greedy or stingy individual.

Questions about complex vocabulary may not be explicitly asked, but this is a useful skill to know. If there is an unfamiliar word while reading a passage and its definition goes unknown, it is possible to miss out on a critical message that could inhibit the ability to appropriately answer the questions. Practicing this technique in daily life will sharpen this ability to derive meanings from context clues with ease.

Practice Questions

Directions: Assume each passage below to be true. Then, pick the answer choice that can be inferred only from the passage itself. Some of the other answer choices might make sense, but only one of them can be derived solely from the passage.

1. Kate has to buy a camera for her trip. She is hiking the Appalachian trail, and it is a requirement that she must pack the lightest equipment possible. Kate has to choose between the compact system camera (CSC) and the digital single-lens reflex camera (DSLR).
   a. Kate has two issues: buying a camera and figuring out which equipment to pack.
   b. Kate will either buy the CSC or the DSLR, depending on which one is smaller.
   c. Kate won't buy the DSLR because it's way too expensive.
   d. Kate is worried that both cameras might be too large to fit in her pack.

2. He adopted a kitten before he went to work. He worried about her all morning. However, he wasn't too concerned when he got a call later that evening about his son being suspended.
   a. He was so worried about the two events he couldn't focus on his work.
   b. His son getting suspended didn't bother him because he was used to it.
   c. The two major events did not occur at the same time.
   d. He worried more about his son being suspended than the kitten.

3. Hard water occurs when rainwater mixes with minerals from rock and soil. Hard water has a high mineral count, including calcium and magnesium. The mineral deposits from hard water can stain hard surfaces in bathrooms and kitchens as well as clog pipes. Hard water can stain dishes, ruin clothes, and reduce the life of any appliances it touches, such as hot water heaters, washing machines, and humidifiers.
   a. Hard water has the ability to reduce the life of a dishwasher.
   b. Hard water is the worst thing to wash your clothes with.
   c. The mineral count in hard water isn't as hard as they say.
   d. Things other than hard water can clog pipes, such as hair and oil.

4. Coaches of kids' sports teams are increasingly concerned about the behavior of parents at games. Parents are screaming and cursing at coaches, officials, players, and other parents. Physical fights have even broken out at games. Parents need to be reminded that coaches are volunteers, giving up their time and energy to help kids develop in their chosen sport. The goal of kids' sports teams is to learn and develop skills, but it's also to have fun. When parents are out of control at games and practices, it takes the fun out of the sport.
   a. Physical fights break out at every single game.
   b. Parents are adding stress to the kids during the game.
   c. Forming a union would help coaches out in their position.
   d. Coaches help kids on sports teams develop their skills and have fun.

5. Tornadoes are dangerous funnel clouds that occur during a large thunderstorm. When warm, humid air near the ground meets cold, dry air from above, a column of the warm air can be drawn up into the clouds. Winds at different altitudes blowing at different speeds make the column of air rotate. As the spinning column of air picks up speed, a funnel cloud is formed. This funnel cloud moves rapidly and haphazardly. Rain and hail inside the cloud cause it to touch down, creating a tornado.

    a. Tornadoes are formed from a mixture of cold and warm air.

    b. Tornadoes are the most dangerous of extreme weather patterns.

    c. Scientists still aren't exactly sure why tornadoes form.

    d. Scientists continue to study tornadoes to improve radar detection and warning times.

6. Digestion begins in the mouth where teeth grind up food and saliva breaks it down, making it easier for the body to absorb. Next, the food moves to the esophagus, and it is pushed into the stomach. The stomach is where food is stored and broken down further by acids and digestive enzymes, preparing it for passage into the intestines.

    a. Food waste is passed into the large intestine.

    b. Nutrients pass into the blood stream while in the small intestine.

    c. As soon as you chew your food, it travels to the esophagus.

    d. Food travels to the esophagus after it is pushed into the stomach.

7. Jordan is the leader of the group, which means she must decide whether the topic for their presentation will be over climate change or food deserts. Jordan is also very diplomatic. Miguel is an expert in climate change while Kennedy has experience growing up in a food desert.

    a. Jordan will pick climate change because it's a more important topic than food deserts.

    b. Jordan has to decide whether personal experience is more important than research experience.

    c. Jordan will flip a coin in order to decide which topic to present over.

    d. Jordan probably likes Kennedy more since she is more relatable, so she will pick her topic.

8. Vacationers looking for a perfect experience should opt out of Disney parks and try a trip on Disney Cruise Lines. While a park offers rides, characters, and show experiences, it also includes long lines, often very hot weather, and enormous crowds.

    a. Although Disney Cruise Lines is fun for the family, it has long lines, very hot weather, and enormous crowds.

    b. Families with small children should not go to Disney parks because there are too many people and the weather is too hot.

    c. At Disney parks, hot weather, lines, and crowds will be less extreme than at Disney Cruise Lines.

    d. At Disney Cruise Lines, hot weather, lines, and crowds will be less extreme than at Disney parks.

9. As summer approaches, drowning incidents will increase. Drowning happens very quickly and silently. Most people assume that drowning is easy to spot, but a person who is drowning doesn't make noise or wave their arms.

    a. Drowning happens silently and decreases as summer approaches.

    b. Drowning happens silently and increases as summer approaches.

    c. Each summer, more children drown than adults.

    d. Many people in summertime wave their arms and make a lot of noise.

10. Last year was the warmest ever recorded in the last 134 years. During that time period, the ten warmest years have all occurred since 2000.
    a. The hottest years in earth's history probably occurred during the dinosaurs' time.
    b. The next 134 years will be hotter than the past ten years.
    c. Out of the last 50 years, the ten warmest years have occurred since 2000.
    d. Burning fossil fuels is what caused the ten warmest years since 2000.

11. A famous children's author recently published a historical fiction novel under a pseudonym; however, it did not sell as many copies as her children's books. In her earlier years, she had majored in history and earned a graduate degree in Antebellum American History, which is the time frame of her new novel.
    a. The author's children's books are more popular than her historical fiction novel.
    b. It's ironic that the author majored in history yet did not sell many copies of her novel.
    c. The author did not sell many copies of her historical fiction novel because it was boring.
    d. Most children's authors cannot cross literary genres without being criticized.

12. Hannah started smoking when she was nineteen years old. The day after Hannah finished a book called *Smoke Free*, she quit. She has been cigarette-free for over a decade.
    a. Hannah will probably have respiratory problems as she gets older.
    b. *Smoke Free* is the reason Hannah quit smoking.
    c. Hannah is at least twenty-nine years old.
    d. Everyone who smokes should read the book *Smoke Free*.

13. Heat loss is proportional to surface area exposed. An elephant loses a great deal more heat than an anteater, because it has a much greater surface area than an anteater.
    a. Surface area causes heat loss.
    b. Too much heat loss can be dangerous.
    c. Elephants lose more heat than anteaters.
    d. Anteaters lose more heat than elephants.

14. The landlord sent an interested tenant the following information about his three apartments for rent: A, B, and C. Apartment B was bigger than Apartment A. Apartment A was in front of Apartment C but above Apartment B. Apartment C was smaller than Apartment A.
    a. Apartment B was above Apartment C.
    b. Apartment B was smaller than Apartment C.
    c. Apartment A was the biggest apartment.
    d. Apartment B was in front of Apartment C.

15. People who argue that William Shakespeare is not responsible for the plays attributed to his name are known as anti-Stratfordians (from the name of Shakespeare's birthplace, Stratford-upon-Avon).
    a. Dr. Porter believes that William Shakespeare is responsible for writing his own plays. He is known as an anti-Stratfordian.
    b. Dr. Filigree believes that William Shakespeare is not responsible for writing his own plays. He is known as an anti-Stratfordian.
    c. Dr. Casings believes that Shakespeare was born somewhere other than Stratford. He is known as an anti-Stratfordian.
    d. Dr. Hendrix believes that Shakespeare died somewhere other than Stratford. He is known as an anti-Stratfordian.

16. Nina is allergic to dairy (which includes cheese and milk), and she doesn't eat any meat except for fish. At his barbecues, Oliver always invites Nina and consistently prepares her a meal that is suitable to her diet. Nina is planning on going to a barbecue later that Oliver is throwing.

    a. Oliver prepares Nina a black bean burger with French fries.

    b. Oliver prepares Nina grilled chicken with asparagus.

    c. Oliver prepares Nina a cheese pizza with a side salad.

    d. Oliver prepares Nina a veggie hot dog with a milkshake.

17. Samuel teaches at a high school in one of the biggest cities in the United States. His students come from diverse family backgrounds. Samuel observes that the best students in his class are from homes where parental supervision is minimal.

a. Samuel should write an academic paper based on his findings.

b. The parents of the bottom five students are probably the most involved.

c. In Samuel's observation, his best students have maximum interference from parents.

d. In Samuel's observation, his best students have minimal interference from parents.

18. Cynthia keeps to a strict vegetarian diet, which is part of her religion. She absolutely cannot have any meat or fish dishes. This is more than a preference; her body has never developed the enzymes to process meat or fish, so she becomes violently ill if she accidentally eats any of the offending foods.

    a. Cynthia doesn't eat meat due to necessity.

    b. Cynthia doesn't eat meat due to preference.

    c. Cynthia doesn't eat meat due to preference as well as necessity.

    d. Cynthia can develop a tolerance to meat by eating small pieces at a time.

19. Samantha wants to be a professional chef, so she started working at a nearby restaurant called *Chesapeake Cuisine*. Samantha also wants to go to college one day to study nutrition. Samantha's mom surprised her later that year by offering to send her to school, but she will only send her if Samantha goes to law school. Samantha cannot afford college on her own.

    a. Samantha's mom is too controlling.

    b. Samantha will never enjoy law school.

    c. Samantha has to choose between going to law school and getting a degree in nutrition.

    d. Samantha has to choose between working at *Chesapeake Cuisine* and going to law school.

20. Barbara had to have an exercise bike for $150 at a store, but soon found out there was a cheaper one online for $75. Barbara always went for cheaper machinery, but hardly ever returned items if she could help it.

    a. Barbara bought both bikes but only used one of them.

    b. Barbara returned the $150 bike and bought the $75 bike.

    c. Barbara kept the $150 bike because she hated returning things.

    d. Barbara did not want a bike that bad so she did not keep any of the bikes.

Answer Explanations

1. B: Kate will either buy the CSC or the DSLR, depending on which one is smaller. We know that Kate "has to buy" a camera for her trip and that "it is a requirement that she must pack the lightest equipment possible." Choice *A* is incorrect. In the passage, we don't see Kate's issue of figuring out which equipment to pack, only which camera to buy. Choice *C* is incorrect. We don't have enough information on the price of the camera to make an educated guess. Choice *D* is incorrect; although we know the camera should be able to fit into Kate's pack, we don't know that Kate is worried about this.

2. C: The two major events did not occur at the same. We know that he worried about the kitten in the morning, and that he found out his son got suspended in the evening. Choice *A* and *D* are incorrect, as the passage states he wasn't too concerned about his son getting suspended. Choice *B* is incorrect; we don't have enough information to make an educated guess about why his son getting suspended didn't bother him.

3. A: Hard water has the ability to reduce the life of a dishwasher. Keep in mind that to make an inference means to make an educated guess based on the facts of the passage. The passage says "hard water can reduce the life of any appliances it touches." Since a dishwasher is an appliance, we can infer that hard water has the ability to reduce the life of a dishwasher. Choice *B* is an opinion and not based on fact. Choice *C* attempts to discredit the passage, and Choice *D* might be true, but we have no evidence of these things in the passage.

4. D: The passage essentially states that coaches help kids on sports teams develop their skills and have fun. Choice *A* is an absolute phrase and is not true in every situation. Choice *B* can be implied, but the passage does not mention the stress of the kids. Choice *C* gives advice beyond the statements in the passage.

5. A: The passage says "when warm, humid air near the ground meets cold, dry air from above, a column of the warm air can be drawn up into the clouds." Thus, we can say that tornadoes are formed from a mixture of cold and warm air. Choice *B* is not necessarily the opinion of the passage. Choices *C* and *D* might be true. However, they are not mentioned in the passage.

6. C: As soon as you chew your food, it travels to the esophagus. Choices *A* and *B* might be correct, but there is no evidence mentioned in the passage. Choice *D* is incorrect; food travels to the stomach after the esophagus, not the other way around.

7. B: Jordan has to decide whether personal experience is more important than research experience. Choice *A* is incorrect; while climate change is a hot topic, we don't know from the passage that it's considered more important than food deserts. Choice *C* is incorrect; although this could happen, it's not the *best* inference of the passage. Choice *D* is incorrect because we know that Jordan is very diplomatic and would not choose a topic in an unfair way. This leaves Choice *B*, which is the best choice because Jordan would be considering the evidence each partner has to offer for the best possible presentation.

8. D: At Disney Cruise Lines, hot weather, lines, and crowds will be less extreme than at Disney parks. Choices *A* and *C* are incorrect and state the opposite sentiment of the passage. Choice *B* might be true, but the passage does not state an opinion of this.

9. B: Drowning happens silently and increases as summer approaches. Choice *A* is incorrect, as this expresses the opposite sentiment. Choice *C* is not mentioned in the passage. Choice *D* uses some of the same language of the passage, but the statement itself is incorrect.

10. C: Out of the last 50 years, the ten warmest years have occurred since 2000. The last 50 years is part of the 134 years that the passage mentions, so this is correct. Choice *A* has no evidence in the passage; neither does Choice *B*. Choice *D* is not mentioned in the passage.

11. A: The author's children's books are more popular than her historical fiction novel. This is expressed by the following statements: "it did not sell as many copies as her children's books." Choices *B, C,* and *D* are not sentiments expressed by the passage.

12. C: Choice A may be true; however, it isn't supported by the text and therefore, it is not the best answer. Choice C is also true and relies on the passage for its information. Choice B is incorrect, as we have no way of knowing what the "deadliest" ingredients are in cigarettes. Finally, Choice D is incorrect; we do not know if "every single chemical" is deadly in a single cigarette, and the passage does not say this.

13. C: Elephants lose more heat than anteaters. The passage states directly that "an elephant loses a great deal more heat than an anteater" because an elephant is larger. We have no way of knowing if Choices *A* or *B* are true according to the passage. Choice *D* is opposite of the correct answer.

14. D: Apartment B was in front of Apartment C. We know this because it states that Apartment A was in front of Apartment C. We also know that Apartment A was on top of Apartment B, which automatically makes Apartment B in front of Apartment C. The rest of the answer choices are logically incorrect based on the information given.

15. B: People known as "anti-Stratfordians" are people who believe that Shakespeare was not responsible for writing his own plays. Choice *B* is the only answer choice that recognizes this fact in the passage. All the other answer choices are incorrect.

16. A: Oliver prepares Nina a black bean burger with French fries. This meal does not include dairy or meat, so this should be an example of Oliver "consistently preparing Nina food that is suitable to her diet." Choice *B* includes meat. Choices *C* and *D* include dairy, cheese pizza and a milkshake.

17. D: In Samuel's observation, his best students have minimal interference from parents. Choices *A* and *B* are suggestions, and not inferences from the text. Choice *C* is opposite of the correct answer.

18. C: Cynthia doesn't eat meat due to preference as well as necessity. Choices *A* and *B* are incorrect because they don't paint the full picture of Cynthia's situation. Choice *D* is incorrect, as this information is not mentioned in the passage.

19. D: The way Samantha's life is set up currently, she has to decide between working at the restaurant and going to law school. Choice *C* is tempting. However, since Samantha cannot afford college on her own, getting a degree in nutrition is simply not an option in this world. Choices *A* and *B* are opinions, and we cannot make an educated guess with the information provided.

20. B: Barbara returned the $150 bike and bought the $75 bike. Look at the language. In the world of the passage, Barbara "always" went for cheaper machinery but "hardly ever" returned items. Therefore, there was a possibility Barbara would return the bike, but there wasn't a possibility she would keep the more expensive bike. With Choices *A* and *D*, we do not have sufficient information to make an educated guess.

Mechanical Comprehension

The *Mechanical Comprehension (MC)* section tests a candidate's knowledge of mechanics and physical principles. These include concepts of force, energy, and work, and how they're used to predict the functioning of tools and machines. This knowledge is important for a successful career in the military. A good score on the MC test shows that a candidate has a solid background for learning how to use tools and machines properly. This is extremely important for the efficient, safe completion of most tasks a future soldier, sailor, or airman must undertake during their service.

The test problems in the MC section of the exam focus on understanding physical principles, but they are *qualitative* in nature rather than *quantitative*. This means the problems involve predicting the *behavior* of a system (such as the direction it moves) rather than calculating a specific measurement (such as its velocity). The figure below shows a sample problem similar to those on the MC test:

Mechanical Comprehension Sample Test Problem

Question 1.

Extending the reach of this crane will shift its

- ○ **A.** total weight
- ○ **B.** allowable speed
- ○ **C.** center of gravity
- ○ **D.** center of buoyancy

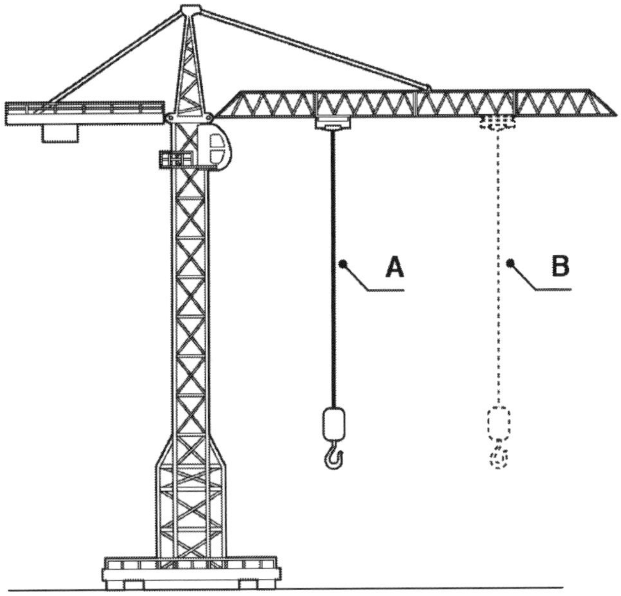

The sample problem pictures a system of a crane lifting a weight, and below the picture is a question. On the exam, it's *very important* to read these questions *carefully*. This question involves completing the following sentence: *Extending the reach of this crane will shift its _____.* After the sentence, four possible answers are provided.

The correct answer is *C, center of gravity*. In this sample problem, it's easy to guess the correct answer simply by eliminating the rest. Answer *A* is incorrect because moving the load out along the crane's boom won't change its weight, just like moving a bodybuilder's arm that's holding a dumbbell won't change the combined weight of the bodybuilder and the dumbbell. Answer *B* is incorrect because the crane isn't moving. That leaves Answers *C* and *D*, but *D* is incorrect because buoyancy is only involved in systems with a liquid (the buoyancy of air is negligible). Therefore, through the process of elimination, *C* is the correct answer.

Review of Physics and Mechanical Principles

The proper use of tools and machinery depends on an understanding of basic physics, which includes the study of motion and the interactions of *mass*, *force*, and *energy*. These terms are used every day, but their exact meanings are difficult to define. In fact, they're usually defined in terms of each other.

The matter in the universe (atoms and molecules) is characterized in terms of its *mass*, which is measured in kilograms in the *International System of Units (SI)*. The amount of mass that occupies a given volume of space is termed *density*.

Mass occupies space, but it's also a component that inversely relates to acceleration when a force is applied to it. This *force* is the application of *energy* to an object with the intent of changing its position (mainly its acceleration).

To understand *acceleration*, it's necessary to relate it to displacement and velocity. The *displacement* of an object is simply the distance it travels. The *velocity* of an object is the distance it travels in a unit of time, such as miles per hour or meters per second:

$$Velocity = \frac{Distance\ Traveled}{Time\ Required}$$

There's often confusion between the words "speed" and "velocity." Velocity includes speed *and* direction. For example, a car traveling east and another traveling west can have the same speed of 30 miles per hour (mph), but their velocities are different. If movement eastward is considered positive, then movement westward is negative. Thus, the eastbound car has a velocity of 30 mph while the westbound car has a velocity of -30 mph.

The fact that velocity has a *magnitude* (speed) and a direction makes it a vector quantity. A *vector* is an arrow pointing in the direction of motion, with its length proportional to its magnitude.

Vectors can be added geometrically as shown below. In this example, a boat is traveling east at 4 *knots* (nautical miles per hour) and there's a current of 3 knots (thus a slow boat and a very fast current). If the boat travels in the same direction as the current, it gets a "lift" from the current and its speed is 7 knots. If the boat heads *into* the current, it has a forward speed of only 1 knot (4 knots – 3 knots = 1 knot) and makes very little headway. As shown in the figure below, the current is flowing north across the boat's path. Thus, for every 4 miles of progress the boat makes eastward, it drifts 3 miles to the north.

Working with Velocity Vectors

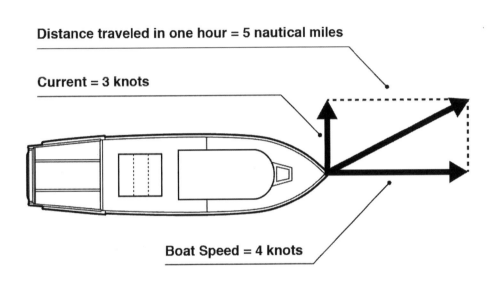

Distance traveled in one hour = 5 nautical miles

Current = 3 knots

Boat Speed = 4 knots

The total distance traveled is calculated using the *Pythagorean Theorem* for a right triangle, which should be memorized as follows:

$$a^2 + b^2 = c^2 \text{ or } c = \sqrt{a^2 + b^2}$$

The problem above was set up using a Pythagorean triple which is made up of positive integers which fit the rule of $a^2 + b^2 = c^2$. In this case, the integers are 3, 4, and 5.

Another example where velocity and speed are different is with a car traveling around a bend in the road. The speed is constant along the road, but the direction (and therefore the velocity) changes continuously.

The *acceleration* of an object is the change in its velocity in a given period of time:

$$Acceleration = \frac{Change\ in\ Velocity}{Time\ Required}$$

91

For example, a car starts at rest and then reaches a velocity of 70 mph in 8 seconds. What's the car's acceleration in feet per second squared? First, the velocity must be converted from miles per hour to feet per second:

$$70\,\frac{miles}{hour} \times \frac{5{,}280\;feet}{mile} \times \frac{hour}{3600\;seconds} = 102.67\;feet/second$$

Starting from rest, the acceleration is:

$$Acceleration = \frac{102.67\,\dfrac{feet}{second} - 0\,\dfrac{feet}{second}}{8\;seconds} = 12.8\;feet/second^2$$

Newton's Laws

Isaac Newton's three laws of motion describe how the acceleration of an object is related to its mass and the forces acting on it. The three laws are:

1. Unless acted on by a force, a body at rest tends to remain at rest; a body in motion tends to remain in motion with a constant velocity and direction.

2. A force that acts on a body accelerates it in the direction of the force. The larger the force, the greater the acceleration; the larger the mass, the greater its inertia (resistance to movement and acceleration).

3. Every force acting on a body is resisted by an equal and opposite force.

To understand Newton's laws, it's necessary to understand forces. These forces can push or pull on a mass, and they have a magnitude and a direction. Forces are represented by a vector, which is the arrow lined up along the direction of the force with its tip at the point of application. The magnitude of the force is represented by the length of the vector.

The figure below shows a mass acted on or "pushed" by two equal forces (shown here by vectors of the same length). Both vectors "push" along the same line through the center of the mass, but in opposite directions. What happens?

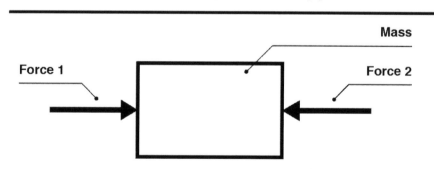

A Mass Acted on by Equal and Opposite Forces

According to Newton's third law, every force on a body is resisted by an equal and opposite force. In the figure above, Force 1 acts on the left side of the mass. The mass pushes back. Force 2 acts on the right side, and the mass pushes back against this force too. The net force on the mass is zero, so according to Newton's first law, there's no change in the *momentum* (the mass times its velocity) of the mass. Therefore, if the mass is at rest before the forces are applied, it remains at rest. If the mass is in motion with a constant velocity, its momentum doesn't change. So, what happens when the net force on the mass isn't zero, as shown in the figure below?

A Mass Acted on by Unbalanced Forces

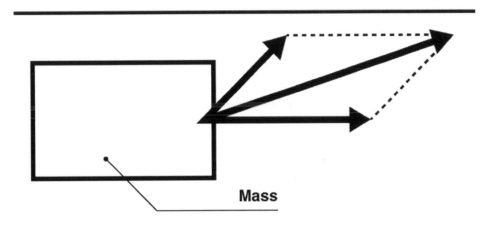

Mass

Notice that the forces are vector quantities and are added geometrically the same way that velocity vectors are manipulated.

Here in the figure above, the mass is pulled by two forces acting to the right, so the mass accelerates in the direction of the net force. This is described by Newton's second law:

Force = Mass x Acceleration

The force (measured in *newtons*) is equal to the product of the mass (measured in kilograms) and its acceleration (measured in meters per second squared or meters per second, per second). A better way to look at the equation is dividing through by the mass:

Acceleration = Force/Mass

This form of the equation makes it easier to see that the acceleration of an object varies directly with the net force applied and inversely with the mass. Thus, as the mass increases, the acceleration is reduced for a given force. To better understand, think of how a baseball accelerates when hit by a bat. Now imagine hitting a cannonball with the same bat and the same force. The cannonball is more massive than the baseball, so it won't accelerate very much when hit by the bat.

In addition to forces acting on a body by touching it, gravity acts as a force at a distance and causes all bodies in the universe to attract each other. The *force of gravity (F_g)* is proportional to the masses of the two objects (*m* and *M*) and inversely proportional to the square of the distance (r^2) between them (and *G* is the proportionality constant). This is shown in the following equation:

$$F_g = G\frac{mM}{r^2}$$

The force of gravity is what causes an object to fall to Earth when dropped from an airplane. Understanding gravity helps explain the difference between mass and weight. Mass is a property of an object that remains the same while it's intact, no matter where it's located. A 10-kilogram cannonball has the same mass on Earth as it does on the moon. On Earth, it *weighs* 98.1 newtons because of the attractive force of gravity, so it accelerates at 9.81 m/s^2. However, on the moon, the same cannonball has a weight of only about 16 newtons. This is because the gravitational attraction on the moon is approximately one-sixth that on Earth. Although Earth still attracts the body on the moon, it's so far away that its force is negligible.

For Americans, there's often confusion when talking about mass because the United States still uses "pounds" as a measurement of weight. In the traditional system used in the United States, the unit of mass is called a *slug*. It's derived by dividing the weight in pounds by the acceleration of gravity (32 feet/s^2); however, it's rarely used today. To avoid future problems, test takers should continue using SI units and *remember to express mass in kilograms and weight in Newtons*.

Another way to understand Newton's second law is to think of it as an object's change in momentum, which is defined as the product of the object's mass and its velocity:

Momentum = Mass x Velocity

Which of the following has the greater momentum: a pitched baseball, a softball, or a bullet fired from a rifle?

A bullet with a mass of 5 grams (0.005 kilograms) is fired from a rifle with a muzzle velocity of 2200 mph. Its momentum is calculated as:

$$2200\,\frac{miles}{hour} \times \frac{5,280\,feet}{mile} \times \frac{m}{3.28\,feet} \times \frac{hour}{3600\,seconds} \times 0.005kg = 4.92\,\frac{kg.m}{seconds}$$

A softball has a mass between 177 grams and 198 grams and is thrown by a college pitcher at 50 miles per hour. Taking an average mass of 188 grams (0.188 kilograms), a softball's momentum is calculated as:

$$50\,\frac{miles}{hour} \times \frac{5280\,feet}{mile} \times \frac{m}{3.28\,ft} \times \frac{hour}{3600\,seconds} \times 0.188kg = 4.19\,\frac{kg.m}{seconds}$$

That's only slightly less than the momentum of the bullet. Although the speed of the softball is considerably less, its mass is much greater than the bullet's.

A professional baseball pitcher can throw a 145-gram baseball at 100 miles per hour. A similar calculation (try doing it!) shows that the pitched hardball has a momentum of about 6.48 kg.m/seconds. That's more momentum than a speeding bullet!

So why is the bullet more harmful than the hard ball? It's because the force that it applies acts on a much smaller area.

Changing the expression of Newton's second law of motion yields a new expression.

$$Force(F) = ma = m \times \frac{\Delta v}{\Delta t}$$

If both sides of the expression are multiplied by the change in time, the law produces the impulse equation.

$$F\Delta t = m\Delta v$$

This equation shows that the amount of force during a length of time creates an impulse. This means that if a force acts on an object during a given amount of time, it will have a determined impulse. However, if the same change in velocity happens over a longer amount of time, the required force is much smaller, due to the conservation of momentum.

$$p = mv$$

In the case of the rifle, the force created by the pressure of the charge's explosion in its shell pushes the bullet, accelerating it until it leaves the barrel of the gun with its *muzzle velocity* (the speed the bullet has when it leaves the muzzle). After leaving the gun, the bullet doesn't accelerate because the gas pressure is exhausted. The bullet travels with a constant velocity in the direction it's fired (ignoring the force exerted against the bullet by friction and drag).

Similarly, the pitcher applies a force to the ball by using their muscles when throwing. Once the ball leaves the pitcher's fingers, it doesn't accelerate and the ball travels toward the batter at a constant speed (again ignoring friction and drag). The speed is constant, but the velocity can change if the ball travels along a curve.

Projectile Motion
According to Newton's first law, if no additional forces act on the bullet or ball, it travels in a straight line. This is also true if the bullet is fired in outer space. However, here on Earth, the force of gravity continues to act so the motion of the bullet or ball is affected.

What happens when a bullet is fired from the top of a hill using a rifle held perfectly horizontal? Ignoring air resistance, its horizontal velocity remains constant at its muzzle velocity. Its vertical velocity (which is zero when it leaves the gun barrel) increases because of gravity's acceleration. Each passing second, the bullet traces out the same distance horizontally while increasing distance vertically (shown in the figure below). In the end, the projectile traces out a *parabolic curve*.

Projectile Path for a Bullet Fired Horizontally from a Hill (Ignoring Air Resistance)

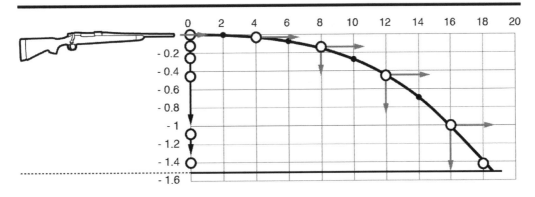

This vertical, downward acceleration is why a pitcher must put an arc on the ball when throwing across home plate. Otherwise the ball will fall at the batter's feet.

It's also interesting to note that if an artillery crew simultaneously drops one cannonball and fires another one horizontally, the two cannonballs will hit the ground at the same time since both balls are accelerating at the same rate and experience the same changes in vertical velocity.

What if air resistance is taken into account? This is best answered by looking at the horizontal and vertical motions separately.

The horizontal velocity is no longer constant because the initial velocity of the projectile is continually reduced by the resistance of the air. This is a complex problem in fluid mechanics, but it's sufficient to note that that the projectile doesn't fly as far before landing as predicted from the simple theory.

The vertical velocity is also reduced by air resistance. However, unlike the horizontal motion where the propelling force is zero after the cannonball is fired, the downward force of gravity acts continuously. The downward velocity increases every second due to the acceleration of gravity. As the velocity increases, the resisting force (called *drag*) increases with the square of the velocity. If the projectile is fired or dropped from a sufficient height, it reaches a terminal velocity such that the upward drag force equals the downward force of gravity. When that occurs, the projectile falls at a constant rate.

This is the same principle that's used for a parachute. Its drag (caused by its shape that scoops up air) is sufficient enough to slow down the fall of the parachutist to a safe velocity, thus avoiding a fatal crash on the ground.

So, what's the bottom line? If the vertical height isn't too great, a real projectile will fall short of the theoretical point of impact. However, if the height of the fall is significant and the drag of the object results in a small terminal fall velocity, then the projectile can go further than the theoretical point of impact.

What if the projectile is launched from a moving platform? In this case, the platform's velocity is added to the projectile's velocity. That's why an object dropped from the mast of a moving ship lands at the base of the mast rather than behind it. However, to an observer on the shore, the object traces out a parabolic arc.

Angular Momentum

In the previous examples, all forces acted through the center of the mass, but what happens if the forces aren't applied through the same line of action, like in the figure below?

A Mass Acted on by Forces Out of Line with Each Other

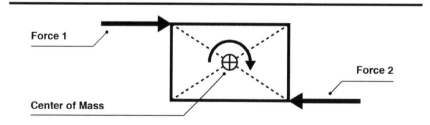

When this happens, the two forces create *torque* and the mass rotates around its center of gravity. In the figure above, the center of gravity is the center of the rectangle ("Center of Mass"), which is determined by the two, intersecting main diagonals. The center of an irregularly shaped object is found by hanging it from two different edges, and the center of gravity is at the intersection of the two "plumb lines."

Newton's second law still applies when the forces form a moment pair, but it must be expressed in terms of angular acceleration and the moment of inertia. The *moment of inertia* is a measure of the body's resistance to rotation, similar to the mass's resistance to linear acceleration. The more compact the body, the less the moment of inertia and the faster it rotates, much like how an ice skater spinning with outstretched arms will speed up as the arms are brought in close to the body.

The concept of torque is important in understanding the use of wrenches and is likely to be on the test. The concept of torque and moment/lever arm will be taken up again below, when the physics of simple machines is presented.

Energy and Work
The previous examples of moving boats, cars, bullets, and baseballs are examples of simple systems that are thought of as particles with forces acting through their center of gravity. They all have one property in common: *energy*. The energy of the system results from the forces acting on it and is considered its ability to do work.

Work or the energy required to do work (which are the same) is calculated as the product of force and distance traveled along the line of action of the force. It's measured in *foot-pounds* in the traditional system (which is still used in workshops and factories) and in *newton meters (N·m)* in the International System of Units (SI), which is the preferred system of measurement today.

Potential and Kinetic Energy
Energy can neither be created nor destroyed, but it can be converted from one form to another. There are many forms of energy, but it's useful to start with mechanical energy and potential energy.

The *potential energy* of an object is equal to the work that's required to lift it from its original elevation to its current elevation. This is calculated as the weight of the object or its downward force (mass times the acceleration of gravity) multiplied by the distance (*y*) it is lifted above the reference elevation or "datum." This is written:

$$PE = mgy$$

The mechanical or *kinetic energy* of a system is related to its mass and velocity and involves the energy of motion:

$$KE = \frac{1}{2}mv^2$$

The *total energy* is the sum of the kinetic energy and the potential energy, both of which are measured in foot-pounds or newton meters.

If a weight with a mass of 10 kilograms is raised up a ladder to a height of 10 meters, it has a potential energy of 10m x 10kg x 9.81m/s^2 = 981N·m. This is approximately 1000 newton meters if the acceleration of gravity (9.81 m/s^2) is rounded to 10 m/s^2, which is accurate enough for most earth-bound calculations. It has zero kinetic energy because it's at rest, with zero velocity.

If the weight is dropped from its perch, it accelerates downward so that its velocity and kinetic energy increase as its potential energy is "used up" or, more precisely, converted to kinetic energy.

When the weight reaches the bottom of the ladder, just before it hits the ground, it has a kinetic energy of 981 N·m (ignoring small losses due to air resistance).

When the 10-kilogram weight hits the ground, its potential energy (which was measured *from* the ground) and its velocity are both zero, so its kinetic energy is also zero. What's happened to the energy? It's dissipated into heat, noise, and kicking up some dust. It's important to remember that energy can neither be created nor destroyed, so it can only change from one form to another.

The conversion between potential and kinetic energy works the same way for a pendulum. If it's raised and held at its highest position, it has maximum potential energy but zero kinetic energy.

Potential and Kinetic Energy for a Swinging Pendulum

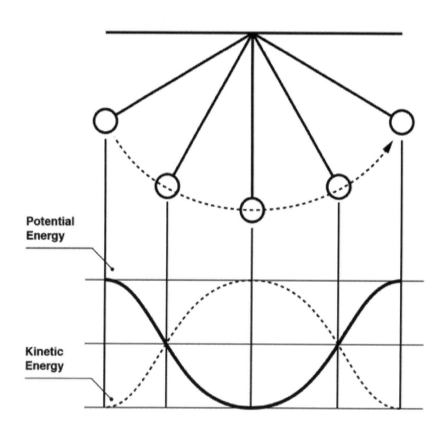

When the pendulum is released from its highest position (see left side of the figure above), it swings down so that its kinetic energy increases as its potential energy decreases. At the bottom of its swing, the pendulum is moving at its maximum velocity with its maximum kinetic energy. As the pendulum swings past the bottom of its path, its velocity slows down as its potential energy increases.

Work
The released potential energy of a system can be used to do *work*.

For instance, most of the energy lost by letting a weight fall freely can be recovered by hooking it up to a pulley to do work by pulling another weight back up (as shown in the figure below).

Using the Energy of a Falling Weight to Raise Another Weight

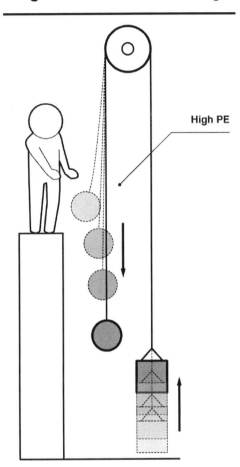

In other words, the potential energy expended to lower the weight is used to do the work of lifting another object. Of course, in a real system, there are losses due to friction. The action of pulleys will be discussed later in this study guide.

Since *energy* is defined as *the capacity to do work*, energy and work are measured in the same units:

$$Energy = Work = Force \times Distance$$

Force is measured in *newtons (N)*. Distance is measured in meters. The units of work are *newton meters (N·m)*. The same is true for kinetic energy and potential energy.

Another way to store energy is to compress a spring. Energy is stored in the spring by stretching or compressing it. The work required to shorten or lengthen the spring is given by the equation:

$$F = k \times d$$

Here, "d" is the length in meters and "k" is the resistance of the spring constant (measured in N×m), which is a constant as long as the spring isn't stretched past its elastic limit. The resistance of the spring is constant, but the force needed to compress the spring increases with each millimeter it's pushed.

The potential energy stored in the spring is equal to the work done to compress it, which is the total force times the change in length.

The potential energy in the spring is stored by locking it into place, and the work energy used to compress it is recovered when the spring is unlocked. It's the same when dropping a weight from a height—the energy doesn't have to be wasted. In the case of the spring, the energy is used to propel an object.

Potential and Kinetic Energy of a Spring

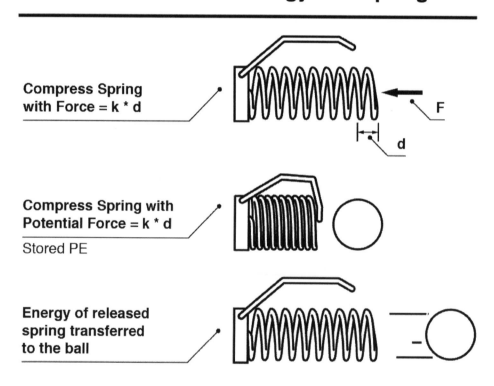

Compress Spring with Force = k * d

Compress Spring with Potential Force = k * d

Stored PE

Energy of released spring transferred to the ball

Pushing a block horizontally along a rough surface requires work. In this example, the work needs to overcome the force of friction, which opposes the direction of the motion and equals the weight of the block times a *friction factor (f)*. The friction factor is greater for rough surfaces than smooth surfaces, and it's usually greater *before* the motion starts than after it has begun to slide. These terms are illustrated in the figure below.

Pushing a Block Horizontally Against the Force of Friction

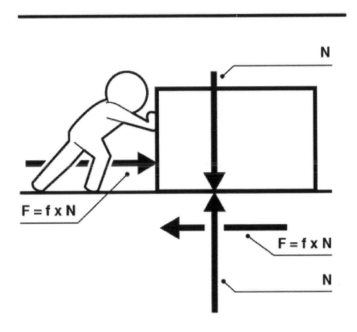

When pushing a block, there's no increase in potential energy since the block's elevation doesn't change. Expending the energy to overcome friction is "wasted" in the generation of heat. Yet, to move a block from point A to point B, an energy cost must be paid. However, friction isn't always a hindrance. In fact, it's the force that makes the motion of a wheel possible.

Heat energy can also be created by burning organic fuels, such as wood, coal, natural gas, and petroleum. All of these are derived from plant matter that's created using solar energy and photosynthesis. The chemical energy or *"heat"* liberated by the combustion of these fuels is used to warm buildings during the winter or even melt metal in a foundry. The heat is also used to generate steam, which can drive engines or turn turbines to generate electric energy.

In fact, work and heat are interchangeable. This fact was first recognized by gun founders when they were boring out cast, brass cannon blanks. The cannon blanks were submerged in a water bath to reduce friction, yet as the boring continued, the water bath boiled away!

Later, the amount of work needed to raise the temperature of water was measured by an English physicist (and brewer) named James Prescott Joule. The way that Joule measured the mechanical equivalent of heat is illustrated in the figure below. This setup is similar to the one in the figure above with the pulley, except instead of lifting another weight, the falling weight's potential energy is converted to the mechanical energy of the rotating vertical shaft. This turns the paddles, which churns the water to increase its temperature. Through a long series of repeated measurements, Joule showed that 4186 N·m of work was necessary to raise the temperature of one kilogram of water by one degree Celsius, no matter how the work was delivered.

Device Measuring the Mechanical Energy Needed to Increase the Temperature of Water

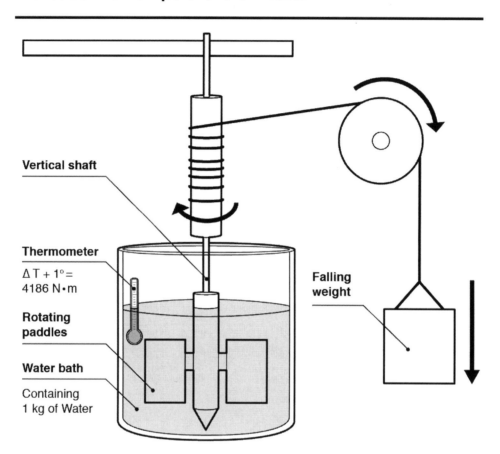

Vertical shaft

Thermometer

ΔT + 1° = 4186 N•m

Rotating paddles

Water bath

Containing 1 kg of Water

Falling weight

In recognition of this experiment, the newton meter is also called a *"joule."* Linking the names for work and heat to the names of two great physicists is truly appropriate because heat and work being interchangeable is of the greatest practical importance. These two men were part of a very small, select group of scientists for whom units of measurement have been named: Marie Curie for radioactivity, Blaise Pascal for pressure, James Watt for power, Andre Ampere for electric current, and only a few others.

Just as mechanical work is converted into heat energy, heat energy is converted into mechanical energy in the reverse process. An example of this is a closely fitting piston supporting a weight and mounted in a cylinder where steam enters from the bottom.

In this example, water is heated into steam in a boiler, and then the steam is drawn off and piped into a cylinder. Steam pressure builds up in the piston, exerting a force in all directions. This is counteracted by the tensile strength of the cylinder; otherwise, it would burst. The pressure also acts on the exposed face of the piston, pushing it upwards against the load (displacing it) and thus doing work.

Work developed from the pressure acting over the area exerts a force on the piston as described in the following equation:

$$Work = Pressure \times Piston\ Area \times Displacement$$

Here, the work is measured in newton meters, the pressure in newtons per square meter or *pascals (Pa)*, and the piston displacement is measured in meters.

Since the volume enclosed between the cylinder and piston increases with the displacement, the work can also be expressed as:

$$Work = Pressure \times \Delta Volume$$

If steam with a pressure slightly greater than this value is piped into the cylinder, it slowly lifts the load. If steam at a much higher pressure is suddenly admitted to the cylinder, it throws the load into the air. This is the principle used to steam-catapult airplanes off the deck of an aircraft carrier.

Power
Power is defined as the rate at which work is done, or the time it takes to do a given amount of work. In the International System of Units (SI), work is measured in *newton meters (N·m)* or *joules (J)*. Power is measured in joules/second or *watts (W)*.

For example, to raise a 1-kilogram mass one meter off the ground, it takes approximately 10 newton meters of work (approximating the gravitational acceleration of 9.81 m/s^2 as 10 m/s^2). To do the work in 10 seconds, it requires 1 watt of power. Doing it in 1 second requires 10 watts of power. Essentially, *doing it faster means dividing by a smaller number*, and that means greater power.

Although SI units are preferred for technical work throughout the world, the old traditional (or English) unit of measuring power is still used. Introduced by *James Watt* (the same man for whom the SI unit of power "watt" is named), the unit of *horsepower (HP)* rated the power of the steam engines that he and his partner (Matthew Boulton) manufactured and sold to mine operators in 18th century England. The mine operators used these engines to pump water out of flooded facilities in the beginning of the Industrial Revolution.

To provide a measurement that the miners would be familiar with, Watt and Boulton referenced the power of their engines with the "power of a horse."

Watt's measurements showed that, on average, a well-harnessed horse could lift a 330-pound weight 100 feet up a well in one minute (330 pounds is the weight of a 40-gallon barrel filled to the brim). Remembering that power is expressed in terms of energy or work per unit time, horsepower came to be measured as:

$$1\ HP = \frac{100\ feet \times 330\ pounds}{1\ minute} \times \frac{1\ minute}{60\ seconds} = 550\ foot\ pounds/second$$

A horse that pulled the weight up faster, or pulled up more weight in the same time, was a more *powerful* horse than Watt's "average horse."

Hundreds of millions of engines of all types have been built since Watt and Boulton started manufacturing their products, and the unit of *horsepower* has been used throughout the world to this day. Of course, modern technicians and engineers still need to convert horsepower to watts to work with SI units. An approximate conversion is *1 HP = 746 W*.

<u>Fluids</u>

In addition to the behavior of solid particles acted on by forces, it is important to understand the behavior of fluids. Fluids include both liquids and gasses. The best way to understand fluid behavior is to contrast it with the behavior of solids, as shown in the figure below.

First, consider a block of ice, which is solid water. If it is set down inside a large box it will exert a force on the bottom of the box due to its weight as shown on the left, in Part A of the figure. The solid block exerts a pressure on the bottom of the box equal to its total weight divided by the area of its base:

$$Pressure = Weight\ of\ block/Area\ of\ base$$

That pressure acts only in the area directly under the block of ice.

If the same mass of ice is melted, it behaves much differently. It still has the same weight as before because its mass hasn't changed. However, the volume has decreased because liquid water molecules are more tightly packed together than ice molecules, which is why ice floats (it is less dense).

The Behavior of Solids and Liquids Compared

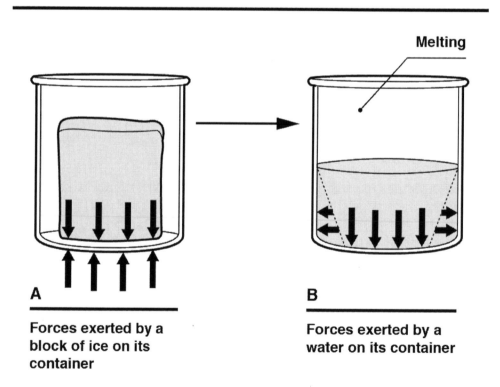

A

Forces exerted by a block of ice on its container

B

Forces exerted by a water on its container

The melted ice (now water) conforms to the shape of the container. This means that the fluid exerts pressure not only on the base, but on the sides of the box at the water line and below. Actually, pressure in a liquid is exerted in all directions, but all the forces in the interior of the fluid cancel each other out, so that a net force is only exerted on the walls. Note also that the pressure on the walls increases with the depth of the water.

The fact that the liquid exerts pressure in all directions is part of the reason some solids float in liquids. Consider the forces acting on a block of wood floating in water, as shown in the figure below.

Floatation of a Block of Wood

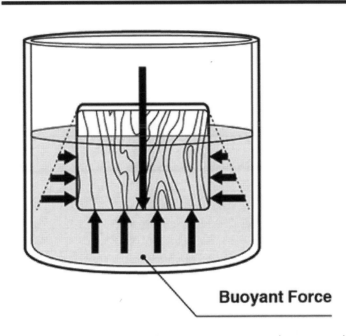

Buoyant Force

The block of wood is submerged in the water and pressure acts on its bottom and sides as shown. The weight of the block tends to force it down into the water. The force of the pressure on the left side of the block just cancels the force of the pressure on the right side.

There is a net upward force on the bottom of the block due to the pressure of the water acting on that surface. This force, which counteracts the weight of the block, is known as the *buoyant force*.

The block will sink to a depth such that the buoyant force of the water (equal to the weight of the volume displaced) just matches the total weight of the block. This will happen if two conditions are met:

1. The body of water is deep enough to float the block
2. The density of the block is less than the density of the water

If the body of water is not deep enough, the water pressure on the bottom side of the block won't be enough to develop a buoyant force equal to the block's weight. The block will be "beached" just like a boat caught at low tide.

If the density of the block is greater than the density of the fluid, the buoyant force acting on the bottom of the boat will not be sufficient to counteract the total weight of the block. That's why a solid steel block will sink in water.

If steel is denser than water, how can a steel ship float? The steel ship floats because it's hollow. The volume of water displaced by its steel shell (hull) is heavier than the entire weight of the ship and its contents (which includes a lot of empty space). In fact, there's so much empty space within a steel ship's hull that it can bob out of the water and be unstable at sea if some of the void spaces (called ballast tanks) aren't filled with water. This provides more weight and balance (or "trim") to the vessel.

The discussion of buoyant forces on solids holds for liquids as well. A less dense liquid can float on a denser liquid if they're *immiscible* (do not mix). For instance, oil can float on water because oil isn't as dense as the water. Fresh water can float on salt water for the same reason.

Pascal's law states that a change in pressure, applied to an enclosed fluid, is transmitted undiminished to every portion of the fluid and to the walls of its containing vessel. This principle is used in the design of hydraulic jacks, as shown in the figure below.

A force (F_1) is exerted on a small "driving" piston, which creates pressure on the hydraulic fluid. This pressure is transmitted through the fluid to a large cylinder. While the pressure is the same everywhere in the oil, the pressure action on the area of the larger cylinder creates a much higher upward force (F_2).

Illustration of a Hydraulic Jack
Exemplifying Pascal's Law

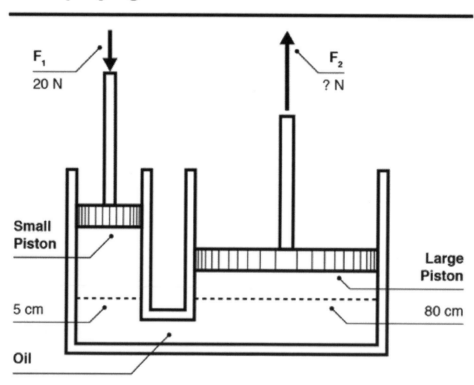

Looking again at the figure above, suppose the diameter of the small cylinder is 5 centimeters and the diameter of the large cylinder is 80 centimeters. If a force of 20 newtons (N) is exerted on the small driving piston, what's the value of the upward force F_2? In other words, what weight can the large piston support?

The pressure within the system is created from the force F_1 acting over the area of the piston:

$$P = \frac{F_1}{A} = \frac{20\ N}{\pi\ (0.05\ m)^2/4} = 10{,}185\ Pa$$

The same pressure acts on the larger piston, creating the upward force, F_2:

$$F_2 = P \times A = 10{,}185\ Pa \times \pi \times (0.8\ m)^2/4 = 5120\ N$$

Because a liquid has no internal shear strength, it can be transported in a pipe or channel between two locations. A fluid's "rate of flow" is the volume of fluid that passes a given location in a given amount of time and is expressed in $m^3/second$. The *flow rate* (Q) is determined by measuring the *area of flow* (A) in m^2, and the *flow velocity* (v) in *m/s*:

$$Q = v \times A$$

This equation is called the *Continuity Equation*. It's one of the most important equations in engineering and should be memorized.

It's important to understand that, for a given flow rate, a smaller pipe requires a higher velocity.

The energy of a flow system is evaluated in terms of potential and kinetic energy, the same way the energy of a falling weight is evaluated. The total energy of a fluid flow system is divided into potential energy of elevation, and pressure and the kinetic energy of velocity. *Bernoulli's Equation* states that, for a constant flow rate, the total energy of the system (divided into components of elevation, pressure, and velocity) remains constant. This is written as:

$$Z + \frac{P}{\rho g} + \frac{v^2}{2g} = Constant$$

Each of the terms in this equation has dimensions of meters. The first term is the *elevation energy*, where Z is the elevation in meters. The second term is the *pressure energy*, where P is the pressure, ρ is the density, and g is the acceleration of gravity. The dimensions of the second term are also in meters. The third term is the *velocity energy*, also expressed in meters.

For a fixed elevation, the equation shows that, as the pressure increases, the velocity decreases. In the other case, as the velocity increases, the pressure decreases.

The use of the Bernoulli Equation is illustrated in the figure below. The total energy is the same at Sections 1 and 2. The area of flow at Section 1 is greater than the area at Section 2. Since the flow rate is the same at each section, the velocity at Point 2 is higher than at Point 1:

$$Q = V_1 \times A_1 = V_2 \times A_2, \qquad V_2 = V_1 \times \frac{A_1}{A_2}$$

Finally, since the total energy is the same at the two sections, the pressure at Point 2 is less than at Point 1. The tubes drawn at Points 1 and 2 would actually have the water levels shown in the figure; the pressure at each point would support a column of water of a height equal to the pressure divided by the unit weight of the water ($h = P/\rho g$).

An Example of Using the Bernoulli Equation

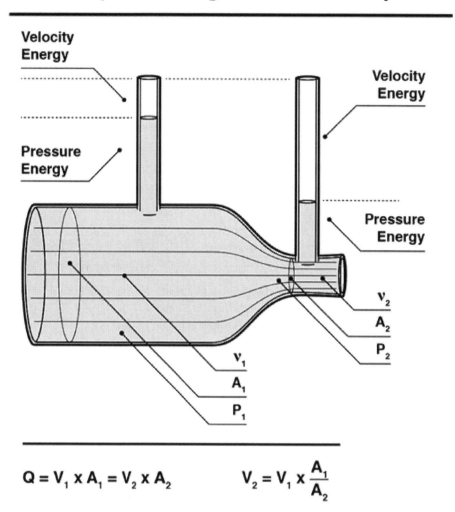

$$Q = V_1 \times A_1 = V_2 \times A_2 \qquad V_2 = V_1 \times \frac{A_1}{A_2}$$

Machines

Now that the basic physics of work and energy have been discussed, the common machines used to do the work can be discussed in more detail.

A *machine* is a device that: transforms energy from one form to another, multiplies the force applied to do work, changes the direction of the resultant force, or increases the speed at which the work is done.

The details of how energy is converted into work by a system are extremely complicated but, no matter how complicated the "linkage" between the components, every system is composed of certain elemental or simple machines. These are discussed briefly in the following sections.

<u>Levers</u>

The simplest machine is a *lever*, which consists of two pieces or components: a *bar* (or beam) and a *fulcrum* (the pivot-point around which motion takes place). As shown below, the *effort* acts at a distance (L_1) from the fulcrum and the *load* acts at a distance (L_2) from the fulcrum.

Components of a Lever

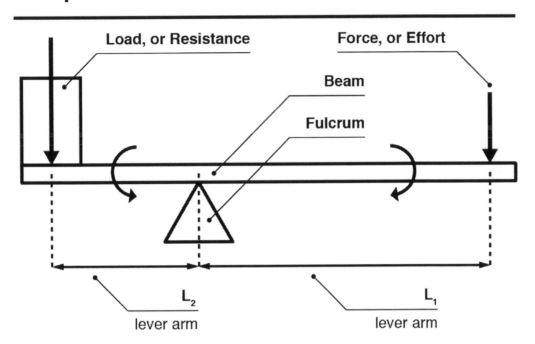

These lengths L_1 and L_2 are called *lever arms*. When the lever is balanced, the load (R) times its lever arm (L_2) equals the effort (F) times its lever arm (L_1). The force needed to lift the load is:

$$F = R \times \frac{L_2}{L_1}$$

This equation shows that as the lever arm L_1 is increased, the force required to lift the resisting load (R) is reduced. This is why Archimedes, one of the leading ancient Greek scientists, said, "Give me a lever long enough, and a place to stand, and I can move the Earth."

The ratio of the moment arms is the so-called "mechanical advantage" of the simple lever; the effort is multiplied by the mechanical advantage. For example, a 100-kilogram mass (a weight of approximately 1000 N) is lifted with a lever like the one in the figure below, with a total length of 3 meters, and the fulcrum situated 50 centimeters from the left end. What's the force needed to balance the load?

$$F = 1000 \ N \times \frac{0.5 \ meters}{2.5 \ meters} = 200 \ N$$

Depending on the location of the load and effort with respect to the fulcrum, three "classes" of lever are recognized. In each case, the forces can be analyzed as described above.

The Three Classes of Levers

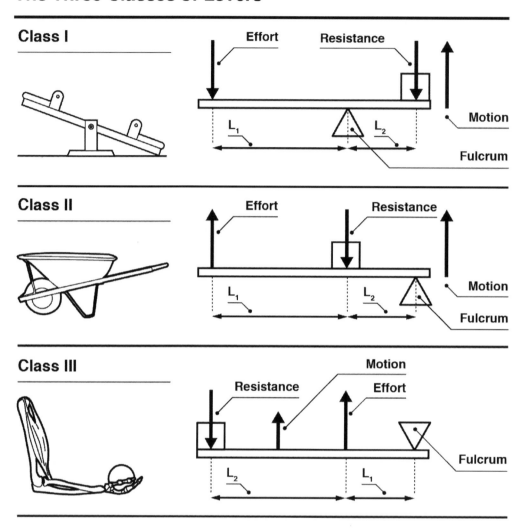

As seen in the figure, a *Class I* lever has the fulcrum positioned between the effort and the load. Examples of Class I levers include see-saws, balance scales, crow bars, and scissors. As explained above, the force needed to balance the load is $F = R \times (L_2/L_1)$, which means that the mechanical advantage is L_2/L_1. The crane boom shown back in the first figure in this section was a Class I lever, where the tower acted as the fulcrum and the counterweight on the left end of the boom provided the effort.

For a *Class II* lever, the load is placed between the fulcrum and the effort. A wheel barrow is a good example of a Class II lever. The mechanical advantage of a Class II lever is $(L_1 + L_2)/L_2$.

For a *Class III* lever, the effort is applied at a point between the fulcrum and the load, which increases the speed at which the load is moved. A human arm is a Class III lever, with the elbow acting as the fulcrum. The mechanical advantage of a Class III lever is $(L_1 + L_2)/L_1$.

<u>Wheels and Axles</u>

The wheel and axle is a special kind of lever. The *axle*, to which the load is applied, is set perpendicular to the *wheel* through its center. Effort is then applied along the rim of the wheel, either with a cable running around the perimeter or with a *crank* set parallel to the axle.

The mechanical advantage of the wheel and axle is provided by the moment arm of the perimeter cable or crank. Using the center of the axle (with a radius of *r*) as the fulcrum, the resistance of the load (*L*) is just balanced by the effort (*F*) times the wheel radius:

$$F \times R = L \times r \quad \text{or} \quad F = L \times \frac{r}{R}$$

This equation shows that increasing the wheel's radius for a given shaft reduces the required effort to carry the load. Of course, the axle must be made of a strong material or it'll be twisted apart by the applied torque. This is why steel axles are used.

<u>Gears, Belts, and Cams</u>

The functioning of a wheel and axle can be modified with the use of gears and belts. *Gears* are used to change the direction or speed of a wheel's motion.

The direction of a wheel's motion can be changed by using *beveled gears*, with the shafts set at right angles to each other, as shown in part *A* in the figure below.

The speed of a wheel can be changed by meshing together *spur gears* with different diameters. A small gear (A) is shown driving a larger gear (B) in the middle section *(B)* in the figure below. The gears rotate in opposite directions; if the driver, Gear A, moves clockwise, then Gear B is driven counter-clockwise. Gear B rotates at half the speed of the driver, Gear A. In general, the change in speed is given by the ratio of the number of teeth in each gear:

$$\frac{Rev_{Gear\,B}}{Rev_{Gear\,A}} = \frac{Number\ of\ Teeth\ in\ A}{Number\ of\ Teeth\ in\ B}$$

Rather than meshing the gears, *belts* are used to connect them as shown in part *(C)*.

Gear and Belt Arrangements

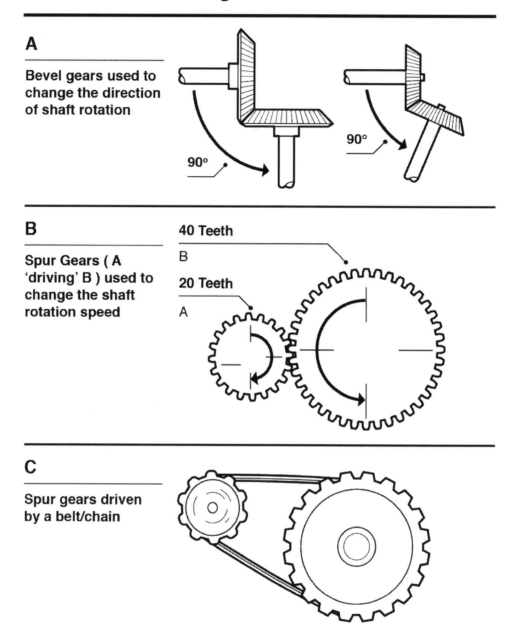

A

Bevel gears used to change the direction of shaft rotation

90°

90°

B

Spur Gears (A 'driving' B) used to change the shaft rotation speed

40 Teeth

B

20 Teeth

A

C

Spur gears driven by a belt/chain

Gears can change the speed and direction of the axle rotation, but the rotary motion is maintained. To convert the rotary motion of a gear train into linear motion, it's necessary to use a *cam* (a type of off-centered wheel shown in the figure below, where rotary shaft motion lifts the valve in a vertical direction.

Conversion of Rotary to Vertical Linear Motion with a Cam

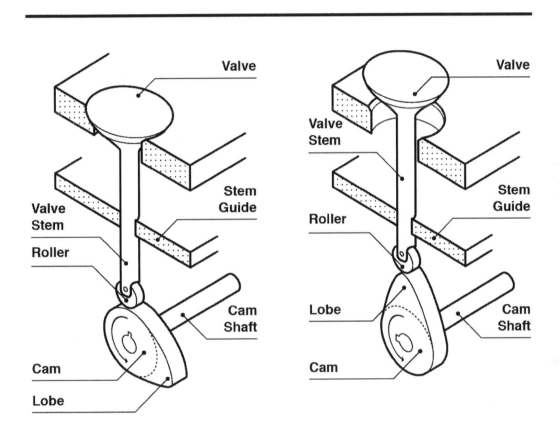

Pulleys

A *pulley* looks like a wheel and axle, but provides a mechanical advantage in a different way. A *fixed pulley* was shown previously as a way to capture the potential energy of a falling weight to do useful work by lifting another weight. As shown in part *A* in the figure below, the fixed pulley is used to change the direction of the downward force exerted by the falling weight, but it doesn't provide any mechanical advantage.

The lever arm of the falling weight (A) is the distance between the rim of the fixed pulley and the center of the axle. This is also the length of the lever arm acting on the rising weight (B), so the ratio of the two arms is 1:0, meaning there's no mechanical advantage. In the case of a wheel and axle, the mechanical advantage is the ratio of the wheel radius to the axle radius.

A *moving pulley*, which is really a Class II lever, provides a mechanical advantage of 2:1 as shown below on the right side of the figure *(B)*.

Fixed-Block Versus Moving-Block Pulleys

A

Single Fixed Block with No Mechanical Advantage

B

Single Moving Block with 2:1 Mechanical Advantage

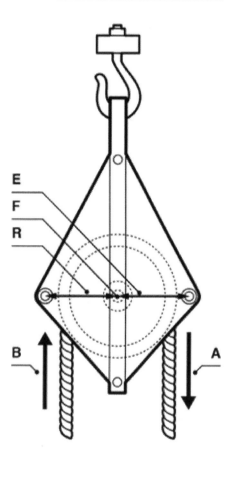

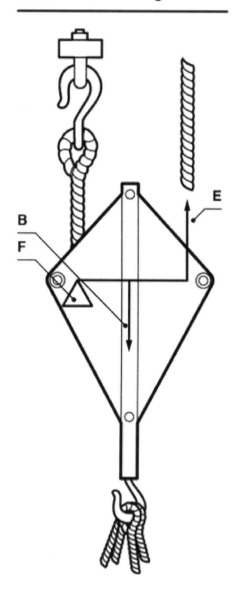

As demonstrated by the rigs in the figure below, using a wider moving block with multiple sheaves can achieve a greater mechanical advantage.

Single-Acting and Double-Acting Block and Tackles

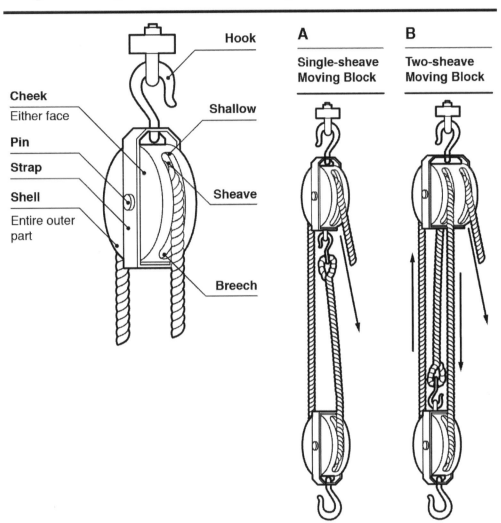

The mechanical advantage of the multiple-sheave block and tackle is approximated by counting the number of ropes going to and from the moving block. For example, there are two ropes connecting the moving block to the fixed block in part A of the figure above, so the mechanical advantage is 2:1. There are three ropes connecting the moving and fixed blocks in part B, so the mechanical advantage is 3:1. The advantage of using a multiple-sheave block is the increased hauling power obtained, but there's a cost; the weight of the moving block must be overcome, and a multiple-sheave block is significantly heavier than a single-sheave block.

Ramps
The *ramp* (or inclined plane) has been used since ancient times to move massive, extremely heavy objects up to higher positions, such as in the pyramids of the Middle East and Central America.

For example, to lift a barrel straight up to a height (*H*) requires a force equal to its weight (*W*). However, the force needed to lift the barrel is reduced by rolling it up a ramp, as shown below. So, if the ramp is *D* meters long and *H* meters high, the force (*F*) required to roll the weight (*W*) up the ramp is:

$$F = \frac{H}{D} \times W$$

Definition Sketch for a Ramp or Inclined Plane

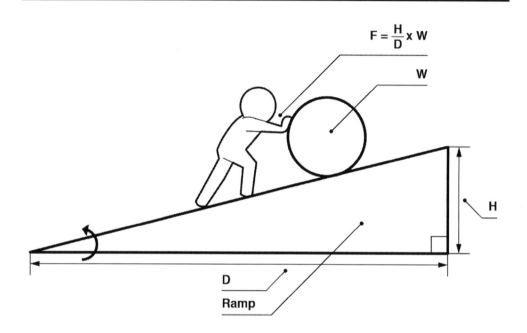

For a fixed height and weight, the longer the ramp, the less force must be applied. Remember, though, that the useful work done (in *N-m*) is the same in either case and is equal to *W* × *H*.

Wedges
If an incline or ramp is imagined as a right triangle like in the figure above, then a *wedge* would be formed by placing two inclines (ramps) back to back (or an isosceles triangle). A wedge is one of the six simple machines and is used to cut or split material. It does this by being driven for its full length into the material being cut. This material is then forced apart by a distance equal to the base of the wedge. Axes, chisels, and knives work on the same principle.

Screws
Screws are used in many applications, including vises and jacks. They are also used to fasten wood and other materials together. A screw is thought of as an inclined plane wrapped around a central cylinder. To visualize this, one can think of a barbershop pole, or cutting the shape of an incline (right triangle) out of a sheet of paper and wrapping it around a pencil (as in part *A* in the figure below). Threads are

made from steel by turning round stock on a lathe and slowly advancing a cutting tool (a wedge) along it, as shown in part *B*.

Definition Sketch for a Screw and Its Use in a Car Jack

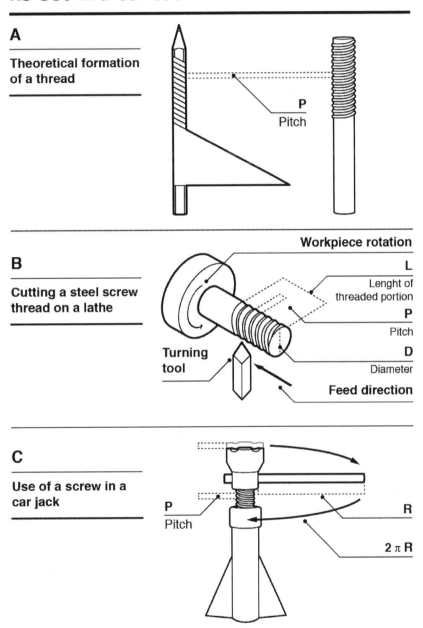

A

Theoretical formation of a thread

P
Pitch

B

Cutting a steel screw thread on a lathe

Workpiece rotation

L
Lenght of threaded portion

P
Pitch

D
Diameter

Turning tool

Feed direction

C

Use of a screw in a car jack

P
Pitch

R

$2\pi R$

The application of a simple screw in a car jack is shown in part *C* in the figure above. The mechanical advantage of the jack is derived from the pitch of the screw winding. Turning the handle of the jack one revolution raises the screw by a height equal to the *screw pitch (p)*. If the handle has a length *R*, the distance the handle travels is equal to the circumference of the circle it traces out. The theoretical mechanical advantage of the jack's screw is:

$$MA = \frac{F}{L} = \frac{p}{2\pi R} \quad \text{so} \quad F = L \times \frac{p}{2\pi R}$$

For example, the theoretical force (F) required to lift a car with a mass (L) of 5000 kilograms, using a jack with a handle 30 centimeters long and a screw pitch of 0.5 cm, is given as:

$$F \cong 50,000 \; N \; \times \; \frac{0.5 \; cm}{6.284 * 30 \; cm} \cong 130 \; N$$

The theoretical value of mechanical advantage doesn't account for friction, so the actual force needed to turn the handle is higher than calculated.

Practice Questions

The following Practice Test contains sample problems that reinforce the principles presented in the *Mechanical Comprehension (MC)* study guide. The answers to these problems, along with a brief explanation, follows.

1. A car is traveling at a constant velocity of 25 m/s. How long does it take the car to travel 45 kilometers in a straight line?
    a. 1 hour
    b. 3600 seconds
    c. 1800 seconds
    d. 900 seconds

2. A ship is traveling due east at a speed of 1 m/s against a current flowing due west at a speed of 0.5 m/s. How far has the ship travelled from its point of departure after two hours?
    a. 1.8 kilometers west of its point of departure
    b. 3.6 kilometers west of its point of departure
    c. 1.8 kilometers east of its point of departure
    d. 3.6 kilometers east of its point of departure

3. A car is driving along a straight stretch of highway at a constant speed of 60 km/hour when the driver slams the gas pedal to the floor, reaching a speed of 132 km/hour in 10 seconds. What's the average acceleration of the car after the engine is floored?
    a. 1 m/s^2
    b. 2 m/s^2
    c. 3 m/s^2
    d. 4 m/s^2

4. A spaceship with a mass of 100,000 kilograms is far away from any planet. To accelerate the craft at the rate of 0.5 m/sec^2, what is the rocket thrust?
    a. 98.1 N
    b. 25,000 N
    c. 50,000 N
    d. 75,000 N

5. The gravitational acceleration on Earth averages 9.81 m/s^2. An astronaut weighs 1962 N on Earth. The diameter of Earth is six times the diameter of its moon. What's the mass of the astronaut on Earth's moon?
    a. 100 kilograms
    b. 200 kilograms
    c. 300 kilograms
    d. 400 kilograms

6. A football is kicked so that it leaves the punter's toe at a horizontal angle of 45 degrees. Ignoring any spin or tumbling, at what point is the upward vertical velocity of the football at a maximum?

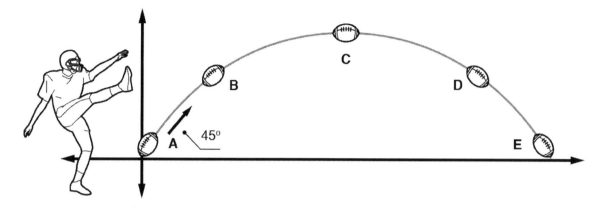

    a. At Point A
    b. At Point C
    c. At Points B and D
    d. At Points A and E

7. The skater is shown spinning in Figure (a), then bringing in her arms in Figure (b). Which sequence accurately describes what happens to her angular velocity?

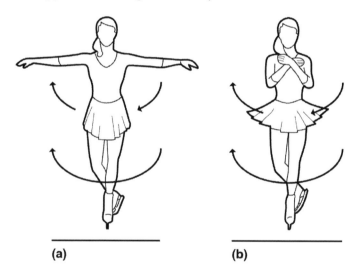

    (a)   **(b)**

    a. Her angular velocity decreases from (a) to (b)
    b. Her angular velocity doesn't change from (a) to (b)
    c. Her angular velocity increases from (a) to (b)
    d. It's not possible to determine what happens to her angular velocity if her weight is unknown.

8. A cannonball is dropped from a height of 10 meters above sea level. What is its approximate velocity just before it hits the ground?
    a. 9.81 m/s
    b. 14 m/s
    c. 32 m/s
    d. It can't be determined without knowing the cannonball's mass

9. The pendulum is held at point A, and then released to swing to the right. At what point does the pendulum have the greatest kinetic energy?

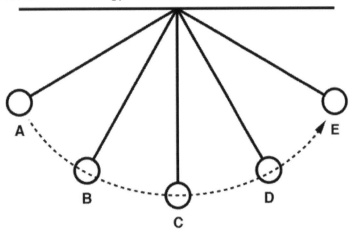

    a. At Point B
    b. At Point C
    c. At Point D
    d. At Point E

10. Which statement is true of the total energy of the pendulum?

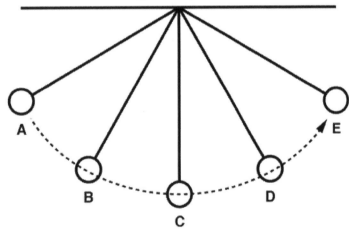

    a. Its total energy is at a maximum and equal at Points A and E.
    b. Its total energy is at a maximum at Point C.
    c. Its total energy is the same at Points A, B, C, D, and E.
    d. The total energy can't be determined without knowing the pendulum's mass.

11. How do you calculate the useful work performed in lifting a 10-kilogram weight from the ground to the top of a 2-meter ladder?
    a. 10kg x 2m x 32 m/s^2
    b. 10kg x 2m^2 x 9.81 m/s
    c. 10kg x 2m x 9.81m/s^2
    d. It can't be determined without knowing the ground elevation

121

12. A steel spring is loaded with a 10-newton weight and is stretched by 0.5 centimeters. What is the deflection if it's loaded with two 10-newton weights?

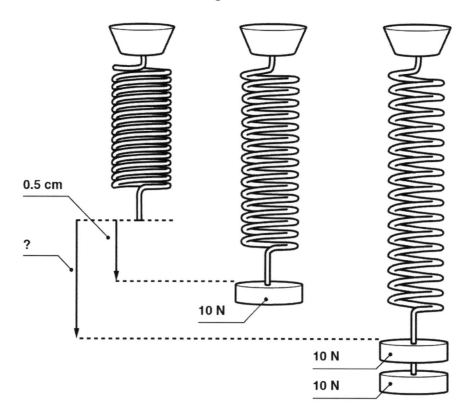

a. 0.5 centimeter
b. 1 centimeter
c. 2 centimeters
d. It can't be determined without knowing the Modulus of Elasticity of the steel.

13. A 1000-kilogram concrete block is resting on a wooden surface. Between these two materials, the coefficient of sliding friction is 0.4 and the coefficient of static friction is 0.5. How much more force is needed to get the block moving than to keep it moving?

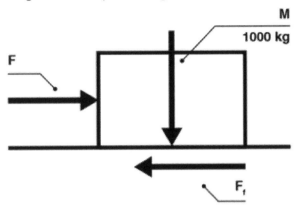

a. 981 N
b. 1962 N
c. 3924 N
d. 9810 N

14. The master cylinder (F1) of a hydraulic jack has a cross-sectional area of 0.1 m^2, and a force of 50 N is applied. What must the area of the drive cylinder (F2) be to support a weight of 800 N?

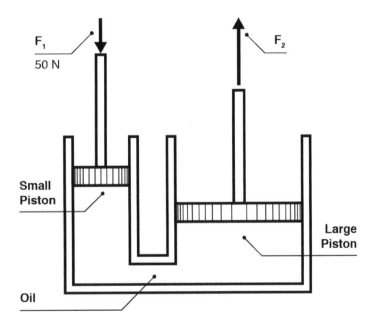

a. 0.4 m^2
b. 0.8 m^2
c. 1.6 m^2
d. 3.2 m^2

15. A gas with a volume V_1 is held down by a piston with a force of F newtons. The piston has an area of A. After heating the gas, it expands against the weight to a volume V_2. What was the work done?

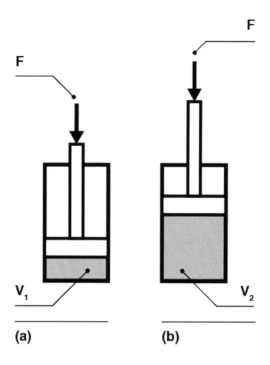

(a)   (b)

a. F/A
b. $(F/A) \times V_1$
c. $(F/A) \times V_2$
d. $(F/A) \times (V_2 - V_1)$

16. A 1000-kilogram weight is raised 30 meters in 10 minutes. What is the approximate power expended in the period?
    a. 1000 Kg × m/s²
    b. 500 N·m
    c. 500 J/s
    d. 100 watts

17. A 2-meter high, concrete block is submerged in a body of water 12 meters deep (as shown). Assuming that the water has a unit weight of 1000 N/m³, what is the pressure acting on the upper surface of the block?

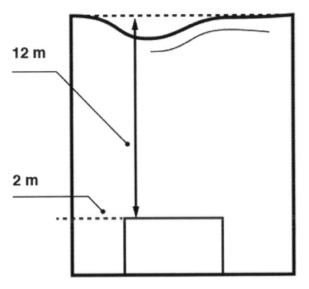

a. 10,000 Pa
b. 12,000 Pa
c. 14,000 Pa
d. It can't be calculated without knowing the top area of the block.

18. Closed Basins A and B each contain a 10,000-ton block of ice. The ice block in Basin A is floating in sea water. The ice block in Basin B is aground on a rock ledge (as shown). When all the ice melts, what happens to the water level in Basin A and Basin B?

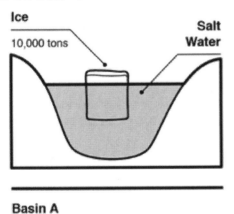

Basin A

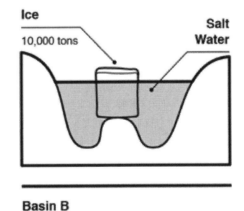

Basin B

a. Water level rises in A but not in B
b. Water level rises in B but not in A
c. Water level rises in neither A nor B
d. Water level rises in both A and B

19. An official 10-lane Olympic pool is 50 meters wide by 25 meters long. How long does it take to fill the pool to the recommended depth of 3 meters using a pump with a 750 liter per second capacity?

    a. 2500 seconds
    b. 5000 seconds
    c. 10,000 seconds
    d. 100,000 seconds

20. Water is flowing in a rectangular canal 10 meters wide by 2 meters deep at a velocity of 3 m/s. The canal is half full. What is the flow rate?

    a. 30 m³/s
    b. 60 m³/s
    c. 90 m³/s
    d. 120 m³/s

21. A 150-kilogram mass is placed on the left side of the lever as shown. What force must be exerted on the right side (in the location shown by the arrow) to balance the weight of this mass?

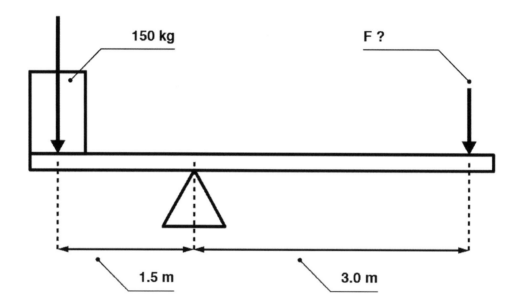

    a. 675 kg.m
    b. 737.75 N
    c. 1471.5 N
    d. 2207.25 N·m

22. For the wheel and axle assembly shown, the shaft radius is 20 millimeters and the wheel radius is 300 millimeters. What's the required effort to lift a 600 N load?

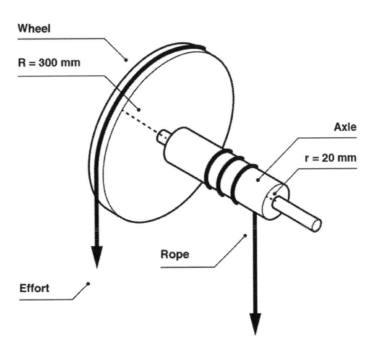

a. 10 N
b. 20 N
c. 30 N
d. 40 N

23. The driver gear (Gear A) turns clockwise at a rate of 60 RPM. In what direction does Gear B turn and at what rotational speed?

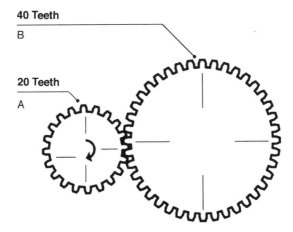

a. Clockwise at 120 RPM
b. Counterclockwise at 120 RPM
c. Clockwise at 30 RPM
d. Counterclockwise at 30 RPM

24. The three steel wheels shown are connected by rubber belts. The two wheels at the top have the same diameter, while the wheel below is twice their diameter. If the driver wheel at the upper left is turning clockwise at 60 RPM, at what speed and in which direction is the large bottom wheel turning?

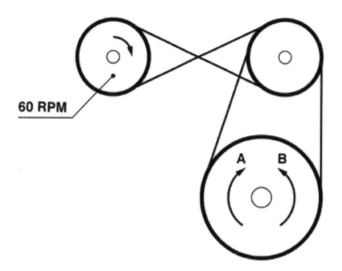

60 RPM

a. 30 RPM, clockwise (A)
b. 30 RPM, counterclockwise (B)
c. 120 RPM, clockwise (A)
d. 120 RPM, counterclockwise (B)

25. In case (a), both blocks are fixed. In case (b), the load is hung from a moveable block. Ignoring friction, what is the required force to move the blocks in both cases?

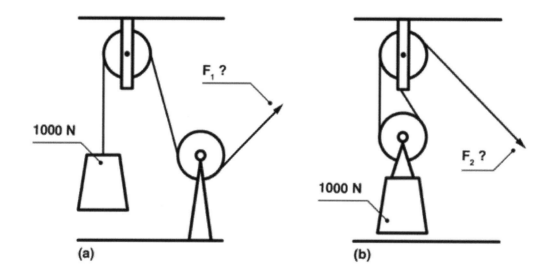

F_1 ?

1000 N

(a)

1000 N

F_2 ?

(b)

a. F_1 = 500 N; F_2 = 500 N
b. F_1 = 500 N; F_2 = 1000 N
c. F_1 = 1000 N; F_2 = 500 N
d. F_1 = 1000 N; F_2 = 1000 N

26. Considering a gas in a closed system, at a constant volume, what will happen to the temperature if the pressure is increased?
   a. The temperature will stay the same
   b. The temperature will decrease
   c. The temperature will increase

27. What is the current when a 3.0 V battery is wired across a lightbulb that has a resistance of 6.0 ohms?
   a. 0.5 A
   b. 18.0 A
   c. 0.5 J

28. According to Newton's Three Laws of Motion, which of the following is true?
   a. Two objects cannot exert a force on each other without touching.
   b. An object at rest has no inertia.
   c. The weight of an object is equal to the mass of an object multiplied by gravity.

29. What is the total mechanical energy of a system?
   a. The total potential energy
   b. The total kinetic energy
   c. Kinetic energy plus potential energy

30. What is the molarity of a solution made by dissolving 4.0 grams of NaCl into enough water to make 120 mL of solution? The atomic mass of Na is 23.0 g/mol and Cl is 35.5 g/mol.
   a. 0.34 M
   b. 0.57 M
   c. 0.034 M

Answer Explanations

1. C: The answer is 1800 seconds:

$$(Desired\ Distance\ in\ km\ \times\ conversion\ factor\ (m\ to\ km)/current\ velocity\ in\ m/s$$

$$\left(45\ km\ \times\ \frac{1000\ m}{km}\right)\Big/25\frac{m}{s} = 1800\ seconds$$

2. D: The answer is 3.6 kilometers east of its point of departure. The ship is traveling faster than the current, so it will be east of the starting location. Its net forward velocity is 0.5 m/s which is 1.8 kilometers/hour, or 3.6 kilometers in two hours.

3. B: The answer is 2 m/s^2:

$$a = \frac{\Delta v}{\Delta t} = \frac{132\frac{km}{hr} - 60\frac{km}{hr}}{10\ seconds} = \frac{70\frac{km}{hr} \times 1000\frac{m}{km} \times \frac{hour}{3600\ sec}}{10\ seconds} = 2\ m/s^2$$

4. C: The answer is 50,000 N. The equation $F = ma$ should be memorized. All of the values are given in the correct units (kilogram-meter-second) so just plug them in.

5. B: The answer is 200 kilograms. This is actually a trick question. The mass of the astronaut is the same everywhere (it is the weight that varies from planet to planet). The astronaut's mass in kilograms is calculated by dividing his weight on Earth by the acceleration of gravity on Earth: 1962/9.81 = 200.

6. A: The answer is that the upward velocity is at a maximum when it leaves the punter's toe. The acceleration due to gravity reduces the upward velocity every moment thereafter. The speed is the same at points A and E, but the velocity is different. At point E, the velocity has a maximum *negative* value.

7. C: The answer is her angular velocity increases from (a) to (b) as she pulls her arms in close to her body and reduces her moment of inertia.

8. B: The answer is 14 m/s. Remember that the cannonball at rest "y" meters off the ground has a potential energy of $PE = mgy$. As it falls, the potential energy is converted to kinetic energy until (at ground level) the kinetic energy is equal to the total original potential energy:

$$\frac{1}{2}mv^2 = mgy \text{ or } v = \sqrt{2gy}$$

This makes sense because all objects fall at the same rate, so the velocity *must* be independent of the mass (which is why "D" is incorrect). Plugging the values into the equation, the result is 14 m/s. Remember, the way to figure this quickly is to have $g = 10$ rather than 9.81.

9. B: The answer is at Point C, the bottom of the arc.

10. C: This question isn't difficult, but it must be read carefully:

A is wrong. Even though the total energy is at a maximum at Points A and E, it isn't equal at only those points. The total energy is the same at *all* points. *B* is wrong. The kinetic energy is at a maximum at C, but not the *total* energy. The correct answer is *C*. The total energy is conserved, so it's the same at *all* points on the arc. *D* is wrong. The motion of a pendulum is independent of the mass. Just like how all objects fall at the same rate, all pendulum bobs swing at the same rate, dependent on the length of the cord.

11. C: The answer is 10kg x 2m x 9.81m/s². This is easy, but it must also be read carefully. Choice *D* is incorrect because it isn't necessary to know the ground elevation. The potential energy is measured *with respect* to the ground and the ground (or datum elevation) can be set to any arbitrary value.

12. B: The answer is 1 centimeter. Remember that the force (*F*) required to stretch a spring a certain amount (*d*) is given by the equation *F = kd*. Therefore, *k = F/d* = 20N/0.5 cm = 20 N/cm. Doubling the weight to 20 N gives the deflection:

$$d = \frac{F}{k} = \frac{20N}{20N/cm} = 1\ centimeter$$

All of the calculations can be bypassed by remembering that the relation between force and deflection is linear. This means that doubling the force doubles the deflection, as long as the spring isn't loaded past its elastic limit.

13. A: The answer is 981 N. The start-up and sliding friction forces are calculated in the same way: normal force (or weight) times the friction coefficient. The difference between the two coefficients is 0.1, so the difference in forces is 0.1 x 1000 x 9.81 = 981 N.

14. C: The answer is 1.6 m². The pressure created by the load is 50N/0.1m² = 500 N/m². This pressure acts throughout the jack, including the large cylinder. Force is pressure times area, so the area equals pressure divided by force or 800N/500N/m² = 1.6m².

15. D: The answer is (*F/A*) x (*V₂ -V₁*). Remember that the work for a piston expanding is pressure multiplied by change in volume. Pressure = *F/A*. Change in volume is (*V₂- V₁*).

16. C: The answer is 500 J/s. Choice *A* is incorrect because kg x m/s² is an expression of force, not power. Choice *B* is incorrect because N·m is an expression of work, not power. That leaves Choices *C* and *D*, both of which are expressed in units of power: watts or joules/second. Using an approximate calculation (as suggested):

$$1000\ kg\ \times\ 10\frac{m}{s^2}\ \times\ 30\ m = 300,000\ N \cdot m\quad so\quad \frac{300,000\ N \cdot m}{600\ seconds} = 500\ watts = 500\ J/s$$

17. B: The answer is 12,000 Pa. The top of the block is under 12 meters of water:

$$P = Unit\ Weight\ of\ Water\ \times\ Depth\ of\ Block\ Under\ Water$$

$$P = 1000\frac{N}{m^3} \times\ 12\ meters = 12,000\frac{N}{m^2} = 12,000\ Pa$$

There are two "red herrings" here: Choice *C* of 14,000 Pa is the pressure acting on the *bottom* of the block (perhaps through the sand on the bottom of the bay). Choice *D* (that it can't be calculated without knowing the top area of the block) is also incorrect. The top area is needed to calculate the total *force* acting on the top of the block, not the pressure.

18. B: The answer is that the water level rises in B but not in A. Why? Because ice is not as dense as water, so a given mass of water has more volume in a solid state than in a liquid state. Thus, it floats. As the mass of ice in Basin A melts, its volume (as a liquid) is reduced. In the end, the water level doesn't change. The ice in Basin B isn't floating. It's perched on high ground in the center of the basin. When it melts, water is added to the basin and the water level rises.

19. B: The answer is 5000 seconds. The volume is 3 x 25 x 50 = 3750 m³. The volume divided by the flow rate gives the time. Since the pump capacity is given in liters per second, it's easier to convert the volume to liters. One thousand liters equals a cubic meter:

$$Time = \frac{Volume}{Flow\ Rate}$$

$$Time = \frac{3,750,000\ liters}{750\ liters/second} = 5000\ seconds = 1.39\ hours$$

20. A: The answer is 30 m³/s. One of the few equations that must be memorized is $Q = vA$. The area of flow is 1m x 10m because only half the depth of the channel is full of water.

21. B: The answer is 737.75 N. This is a simple calculation:

$$acceleration\ due\ to\ gravity \times mass\ of\ object\ in\ kg\ \times distance\ to\ fulcrum$$

$$\frac{9.81\ m}{s^2} \times 150\ kg\ \times 1.5\ m = 3\ m\ \times F \quad so \quad F = \frac{2207.25\ N \cdot m}{3\ meters}$$

21. A 150-kilogram mass is placed on the left side of the lever as shown. What force must be exerted on the right side (in the location shown by the arrow) to balance the weight of this mass?

22. D: The answer is 40 N. Use the equation $F = L \times r/R$. Note that for an axle with a given, set radius, the larger the radius of the wheel, the greater the mechanical advantage.

23. D: The answer is counterclockwise at 30 RPM. The driver gear is turning clockwise, and the gear meshed with it turns counter to it. Because of the 2:1 gear ratio, every revolution of the driver gear causes half a revolution of the follower.

24. B: The answer is 30 RPM, counterclockwise (B). While meshed gears rotate in different directions, wheels linked by a belt turn in the same direction. This is true unless the belt is twisted, in which case they rotate in opposite directions. So, the twisted link between the upper two wheels causes the right-hand wheel to turn counterclockwise, and the bigger wheel at the bottom also rotates counterclockwise. Since it's twice as large as the upper wheel, it rotates with half the RPMs.

25. C: The answer is F_1 = 1000 N; F_2 = 500 N. In case (a), the fixed wheels only serve to change direction. They offer no mechanical advantage because the lever arm on each side of the axle is the same. In case (b), the lower moveable block provides a 2:1 mechanical advantage. A quick method for calculating the mechanical advantage is to count the number of lines supporting the moving block (there are two in this question). Note that there are no moving blocks in case (a).

26. C: According to the *ideal gas law* ($PV = nRT$), if volume is constant, the temperature is directly related to the pressure in a system. Therefore, if the pressure increases, the temperature will increase in direct proportion. Choice *A* would not be possible, since the system is closed and a change is occurring, so the temperature will change. Choice *B* incorrectly exhibits an inverse relationship between pressure and temperature, or $P = 1/T$.

27. A: According to Ohm's Law: $V = IR$, so using the given variables: $3.0\ V = I \times 6.0\ \Omega$

Solving for I: $I = 3.0\ V/6.0\ \Omega = 0.5\ A$

Choice *B* incorporates a miscalculation in the equation by multiplying 3.0 V by 6.0 Ω, rather than dividing these values. Choices *C* is labeled with the wrong units; Joules measure energy, not current.

28. C: The weight of an object is equal to the mass of the object multiplied by gravity. According to Newton's Second Law of Motion, $F = m \times a$. Weight is the force resulting from a given situation, so the mass of the object needs to be multiplied by the acceleration of gravity on Earth: $W = m \times g$. Choice *A* is incorrect because, according to Newton's first law, all objects exert some force on each other, based on their distance from each other and their masses. This is seen in planets, which affect each other's paths and those of their moons. Choice *B* is incorrect because an object in motion or at rest can have inertia; inertia is the resistance of a physical object to change its state of motion.

29. C: In any system, the total mechanical energy is the sum of the potential energy and the kinetic energy. Either value could be zero but it still must be included in the total. Choices *A* and *B* only give the total potential or kinetic energy, respectively.

30. B: To solve this, the number of moles of NaCl needs to be calculated:

First, to find the mass of NaCl, the mass of each of the molecule's atoms is added together as follows:

$$23.0g\ (Na) + 35.5g\ (Cl) = 58.8g\ NaCl$$

Next, the given mass of the substance is multiplied by one mole per total mass of the substance:

$$4.0g\ NaCl \times (1\ mol\ NaCl/58.5g\ NaCl) = 0.068\ mol\ NaCl$$

Finally, the moles are divided by the number of liters of the solution to find the molarity:

$$(0.068\ mol\ NaCl)/(0.120L) = 0.57\ M\ NaCl$$

Choice *A* incorporates a miscalculation for the molar mass of NaCl, and Choice *C* incorporates a miscalculation by not converting mL into liters (L), so it is incorrect by a factor of 10.

Dear OAR Test Taker,

We would like to start by thanking you for purchasing this study guide for your OAR exam. We hope that we exceeded your expectations.

Our goal in creating this study guide was to cover all of the topics that you will see on the test. We also strove to make our practice questions as similar as possible to what you will encounter on test day. With that being said, if you found something that you feel was not up to your standards, please send us an email and let us know.

We would also like to let you know about other books in our catalog that may interest you.

ASVAB

This can be found on Amazon: amazon.com/dp/1628454970

ASTB

amazon.com/dp/162845508X

AFOQT

amazon.com/dp/1628454776

SIFT

amazon.com/dp/1628454318

We have study guides in a wide variety of fields. If the one you are looking for isn't listed above, then try searching for it on Amazon or send us an email.

Thanks Again and Happy Testing!
Product Development Team
info@studyguideteam.com

Interested in buying more than 10 copies of our product? Contact us about bulk discounts:

bulkorders@studyguideteam.com

FREE Test Taking Tips DVD Offer

To help us better serve you, we have developed a Test Taking Tips DVD that we would like to give you for FREE. **This DVD covers world-class test taking tips that you can use to be even more successful when you are taking your test.**

All that we ask is that you email us your feedback about your study guide. Please let us know what you thought about it – whether that is good, bad or indifferent.

To get your **FREE Test Taking Tips DVD**, email freedvd@studyguideteam.com with "FREE DVD" in the subject line and the following information in the body of the email:

   a. The title of your study guide.

   b. Your product rating on a scale of 1-5, with 5 being the highest rating.

   c. Your feedback about the study guide. What did you think of it?

   d. Your full name and shipping address to send your free DVD.

If you have any questions or concerns, please don't hesitate to contact us at freedvd@studyguideteam.com.

Thanks again!

Made in the USA
Columbia, SC
14 April 2019